Plant Pathology

NIPA® GENX ELECTRONIC RESOURCES & SOLUTIONS P. LTD.
New Delhi-110 034

About the Author

Dr. Sanjeev Kumar is Assistant professor/Scientist, Department of Plant Pathology, Jawaharlal Nehru Krishi Vishwavidyalaya, Jabalpur, Madhya Pradesh. He is very sincere and hard working teacher. He has taught many undergraduate, postgraduate and Ph.D courses with great dedication to the great satisfaction of the students and actively involved in all teaching activities. Dr. Kumar has guided nineteen M.Sc and two Ph.D student and has been actively involved in his research and extension activities too. As a researcher, he has been associated with 8 major research projects sponsored by ICAR, NATP,TSP and JICA, Japan. He has published 68 research and reviews papers, 25 book chapters, 105 popular articles, 1 research bulletin and 3 technical folders. He is an active member of over six National and International Societies. He is the fellow member of Indian Phytopathological Society, IARI New Delhi, Indian Society of Mycology and Plant Pathology, Udaipur, Rajasthan and Society of Biocontrol Advancement, Banguluru. He has attended many conferences, symposia, workshops and presented papers. He has also attended the World Soybean Research Conference held in Durban, South Africa.

Dr. Kumar has authored nine books entitled Plant Pathogens & Principles of Plant Pathology, Diseases of Horticultural Crops: Identification & Management, Diseases of Field Crops and their Integrated Management, Pesticides and Plant Protection Appliances, Fundamentals of Plant Pathology, Integrated Disease Management: Principles & Practices, Principles of Plant Pathology published by New India Publishing Agency, Pitam Pura, New Delhi, Diseases of Field & Horticultural Crops & Their Management-I, published by Brillion Publishing, DeshBandhu Gupta Road, Karol Bagh, New Delhi-110005 and Principles of Plant Disease Management Diseases, published by Kalyani Publishers, Daryaganj ,New Delhi and three practical manuals. His research interests include Pulse pathology and biological control

---

for other volumes of
The Professional's Notes Series

Plant Pathology
The Professional's Notes Series
Volume 06

Sanjeev Kumar
Assistant Professor/Scientist
Department of Plant Pathology
Office of Dean, Faculty of Agriculture
Jawahar Lal Nehru Krishi Vishwavidyalaya
Jabalpur-482004, M.P., India

NIPA® GENX ELECTRONIC RESOURCES & SOLUTIONS P. LTD.
New Delhi-110 034

NIPA® GENX ELECTRONIC RESOURCES & SOLUTIONS P. LTD.

101,103, Vikas Surya Plaza, CU Block
L.S.C.Market, Pitam Pura, New Delhi-110 034
Ph : +91 11 27341616, 27341717, 27341718
E-mail: newindiapublishingagency@gmail.com
www: www.nipabooks.com

For customer assistance, please contact
Phone: + 91-11-27 34 17 17
Fax: + 91-11- 27 34 16 16

Print ISBN: 978-93-58870-73-2
ebook ISBN: 978-93-58872-99-6

Composed and Designed by NIPA®.

Preface

The current book "Plant Pathology" is an attempt to provide graduate and post-graduate students with a tool that will help them supplement their efforts for competitive examinations for entrance to various degree courses and preparation for JRF/SRF/NET/ARS. Discussions with learners and educators revealed that a practical, high-quality plant pathology book for competitive preparation had long been desired. As a result, In nine units, the book covers the history and principles of plant pathology, laboratory and analytical techniques, physiological and molecular plant pathology, mycology & classification of fungi, plant bacteriology, plant virology, plant disease epidemiology, phanerogamic parasites and non-parasitic diseases, and plant disease management. The book follows ICAR's syllabus for JRF/SRF/NET/ARS exams. Each chapter is written point by point to inspire individual thought and critical memory assessment in order to retain the chapter's key themes.

I attempted to synthesize complete knowledge on plant pathology by gathering and digesting all relevant information published in reputable phytopathological and associated biological publications, specialized books, annual reviews and online. This is a compilation work that provides a pretty complete and concise study of the fundamental facts and elements of plant pathology to help students succeed in competitive examinations.

We hope that students and readers who enjoy working in the subject of plant pathology will find the book's format, level, and quantity of information suitable for easier understanding and learning. The broad spectrum of information is provided briefly and point-by-point in separate chapters. The readers are anticipated to find the structure engaging and helpful to learning and confidently preparing for competitive tests.

This book is dedicated to my parents, who gave up everything to provide me with the best education possible; to my wife, "Archana," whose love and support have been the most priceless things in our life together; and, lastly, to our youngest children, "Saumya" and "Adyan," who one day will be able to read their names in the book and feel proud of their father and reassured of "Papa's" love for them.

I wish to express my sincere gratitude to our past and present pioneers in the field of Plant Pathology whose work has been freely used in the creation of this book.

Your useful feedback and suggestions are welcome to help us improve it further.

Jabalpur **Sanjeev Kumar**

Contents

Unit 3: Physiological and Molecular Plant Pathology

Unit 4: Mycology Classification of Fungi

Unit 5: Plant Bacteriology

Unit 6: Plant Virology

Unit 7: Plant Disease Epidemiology

Unit 8: Phanerogamic Parasites and Non-parasitic Diseases

Unit 9: Management of Plant Diseases

Unit 1: History and Principles of Plant Pathology

1

Plant Pathology Definition, Objective, Scope and Importance of Plant Pathology

Plant Pathology

Plant Pathology is derived from Greek words- *pathos* (suffering) + *Logos* (study) Study of the suffering plant.

Two phases

- Science: *Understanding of the disease* i.e. *how do pathogens invade, how do plants defend themselves, what causes symptoms* etc.
- Art: *Application of the science to the field problems* i.e. Diagnosis and Control.

Defnition

Plant pathology or phytopathology is the branch of agricultural, science which deals with the cause, etiology, resulting in losses and management methods of plant diseases.

What Plant Pathology Tries to Do, As A Science ?

- To increase our knowledge about plant diseases.
- To develop methods, equipment and materials through which plant diseases can be avoided or controlled.
- Controlling plant disease results in more food of better quality and a more aesthetically pleasing environment.
- To help in controlling food prices.
- To help in reducing diseased plant produce which may sometimes be poisonous and unfit for consumption.
- To manage some plant diseases which wipe out entire plant species and many affect the beauty and landscape of our environment.

Plant Pathology: An Integrative Science and Profession

- Uses and combines the basic knowledge of Botany, Mycology, Bacteriology, Virology, Nematology, Plant Anatomy, Plant Physiology,

Genetics, Molecular Biology and Genetic Engineering, Biochemistry, Horticulture, Agronomy, Tissue Culture, Soil Science, Forestry, Chemistry, Physics, Meteorology, and many other branches of science.

- Plant pathology profits from advances in any one of these sciences, and many advances in other sciences have been made in attempts to solve plant pathological problems.

Major Objectives

- To study biotic, mesobiotic and abiotic causes of diseases or disorders.
- To study the mechanisms of disease development by pathogens.
- To study the plant (host)-pathogen interaction in relation to environment.
- To develop methods of management of plant diseases.

Scope of Plant Pathology

- Survey of crop fields to find out prevalence of diseases and their incidence.
- Recording new diseases of economic importance if any, with their identification and extent of incidence.
- Assessment of losses caused by different diseases of economic importance.
- Study of etiology, symptoms, predisposing factors and recurrence of such diseases.
- Find out suitable and economic methods of management of economically important plant diseases.
- Assisting in breeding of disease resistant varieties.
- To train the extension workers and subject matter specialists in order to bridge the gap between pathologists and farmers for better crop production.

Importance of Plant Pathology in Agriculture

1. Advanced the knowledge to protect the crop from losses due to diseases.
2. Most of the diseases with known disease cycle can now be avoided by the modification of cultural practices.
3. With the knowledge of mode of disease spread, many diseases of economic importance can now be managed.
4. Crop improvement and varietal resistance have been achieved against many diseases through the joint effort of breeder and plant pathologist.
5. Contributed in disease free certified seed production.

6. Made possible to restrict the spread of plant diseases from one place to other and one country to other through suitable measures and quarantine legislation.
7. Various prophylactic measures like seed treatment, soil treatment and crop rotations are adopted for successful management of diseases.

Challenges for Plant Pathology

- To reduce food losses.
- To improve food quality.
- To safeguard our environment.
- To control plant disease effectively and safely .
- To feed billions of hungry people.

Most promising approaches of Controlling Plant Diseases

1. Conventional breeding and genetic engineering of disease-resistant plants.
2. Application of disease-suppressing cultural practices.
3. RNA and gene-silencing techniques of plant defense promoting, nontoxic substances.
4. Use of biological agents antagonistic to the microorganisms that cause plant disease.

2

Milestones in Phytopathology

Historical Developments of Plant Pathology; Mycology

Year	Contributor (s)	Description
1885	KR Kritikar	First Indian scientist who collected and identified fungi.
1886-1971	JF Dastur	First Indian plant pathologist known for the establishment of genus Phytophthora and diseases caused by it in castor and potato.
	TS Sadasivan	Developed concept of vivotoxins and worked out the mechanism of wilting in cotton owing to *Fusarium oxysporum* f.sp. *vasinfectum.*
	SN Dasgupta	Carried out extensive study on black tip of mango.
	MJ Thrimalachar	Discovered several antifungal antibiotics *viz* aureofungin, haymycin etc.
	YL Nene	Reported Khaira disease of rice caused due to Zn deficiency. Authored book "*Fungicides in Plant Disease Control*"
1889	DD Cunningham	Reported the causal organism of red rust of tea in Assam caused by *Cephaleurous virescens*
1920	EJ Butler	First direcotor of Imperial Mycological Institue in England. Initiated an exhaustive study on Indian fungi and the disease caused by them. Discovered genus Allomyces. Wrote "*Fungi and Diseases in Plants*". Father of Indian Plant Pathology
1940	KC Mehta	Studied epidemiology of cereal rusts in IndiaWrite monograph on "*Further studies on cereal rust in India*" Father of Indian Rust
1948	BB Mundukar	Established Indian Phytopathological society with its journal "*Indian Phytopathology*". Book; Fungi and Plant diseases
1953	JC Luthra and A Sattar	Developed solar heat treatment for loose smut of wheat
	Ramanujam and Co-workers	Developed seed plot technique for virus free seed potato production in Indo Gangetic plains
1975	S Nagrajan and H Singh	Formulated Indian Stem Rust Rules for *Puccinia graminis tritici*
1885	PM Millardet	Discovered the *Bordeaux mixture* for the control of downey mildew of grapes

1885	Frank	Discovered *Mycorrhizal* fungi
1887	Mason	Introduced *Burgundy mixture*
1887	Jensen	Developed *hot water technique* for wheat smut.
1894	Ericsson	Reported *Physiological specialization* in stem rust of wheat
1904	AF Blakeslee	Founded *heterothallism* in *Rhizopus*
1905	RH Biffen	Demonstrated inheritance of rust resistance in wheat in a Medelain fashion
1917	EC Stakman	Distinguished *biological forms* in cereal rusts
1923	Hansen and Smith	Term *Heterokaryosis*
1931	PA Saccardo	Wrote "*Syllogue Fungorum*"
1931	JC Luthra	*Solar heat treatment* for loose smut of wheat.
1946	HH Flor	Gave *gene for gene hypothesis* while working on linseed rust
1952	G Pontecorvo and JA Roper	Discovered *parasexuality* in *Aspergillus nidulans*
1968	Vander Plank	Concept of horizontal and vertical resistance. Father of Plant Epidemiology
1990	DF Klessig and I Raskin; JP Metraux and J Tyals	Demonstrated that *salicylic acid* is associated with systemic acquired resistance
2005	RA Dean and Co-workers	The first complete genome sequencing of a plant pathogenic fungus *Magnaporthe grisea*

Historical Developments of Plant Pathology

1. Plant Bacteriology

Year	Contributor (s)	Description
1683	Anton von Leeuwenhoek	First observed bacteria
1876	Robert Koch	They proved that anthrax disease of cattle was caused by specific bacterium
1876	Louis Pasteur	Demonstrated role of bacteria in fermentation and decay
1876	Robert Koch	Described the theory called "Koch's postulates." He established the principles of pure culture technique
1876	Robert Koch and Pasteur	Disproved the theory of spontaneous generation of diseases & propose germ theory in relation to the diseases of man and animal
1876	Woronin	Isolated and described the root nodule bacteria in leguminous plant
1878	TJ Burril	Described first bacterial disease (Fire blight of apples) caused by *Erwinia amylovora*

1887	EF Smith	Gave the final proof that bacteria could be the incitants of plant diseases. First to notice and study the crown gall disease. Father of Phytobacteriology
1980	DW Dye and Co-workers	Introduced pathovar system in taxonomy
1887	EF Smith	Gave the final proof that bacteria could be the incitants of plant diseases. First to notice and study the crown gall disease. Father of Phytobacteriology
1980	DW Dye and Co-workers	Introduced pathovar system in taxonomy
1952	J Lederberg	Coined the term plasmid
1952	SA Waksman	Discovered streptomycin
1952	Zinder and J Lederberg	Discovered transduction in bacteria
1962	H Stolp	Discovered bdellovibrios
1972	PB New and A Kerr	Success in biological control of *A. radiobacter* strain K.
1974	I Zanen et al.	Demonstrated Ti plasmid in *Agrobacterium tumefaciens*

2. Virology

Year	Contributor (s)	Description
1576	C Clusius	Recorded breaking of tulips owing to viral infection
1886	AE Mayer	First to point out that *tobacco mosaic is readily transmissible and infectious.* Demonstrated the sap transmission of the TMV
1892	D Ivanowski	Demonstrated that the agents of *TMV could pass through filters that retained bacterial cells* (bacteria proof filter)
1898	MW Beijernick	Concept of *contagium vivum fluidum.* Father of Plant Virology
1917	d'Herelle	Discovered *bacteriophage*
1932	Knoll and Ruska	Invented Electron microscope
1935	WM Stanley	Crystallized *Tobacco mosaic virus*
1936	Bawden & NW Pirie	Found that the crystalline nature of the virus contains *nucleic acid and protein*
1956	Gierer & Schramm	Proved that the nucleic acid fraction of the virus is actually the infectious agent
1966	Kassanis	Discovered the *satellite viruses*
1963 & 1966	Black & Markhan; Miura et al.	Showed that wound tumour and rice dwarf viruses contain double started RNA.

3. Plant Disease Control

Year	Contributor (s)	Description
1000	Homer	Mentioned about the use of sulphur in plant disease control

1807	B Prevost	Recommended copper sulphate for wheat seed treatment against bunt and demonstrated first time the fungitoxic value of copper compounds
1882	PMA Millardet	Discovered Bordeaux mixture for control of downy mildew of grape vine
1887	Mason	Discovered burgundy mixture
1921	Bewley	Developed chestnut compound
1934	Tisdale & Williams	Reported fungitoxicity of dithiocarbamates
1942	Singh	Delveloped Chaubatia paste
1943	SA Waksman & A Schatx	Discovered Streptomycin
1952	Kittleson	Discovered captan also known as Kittleson's Killer
1966	Von Schmeling and M Kulka	Discovered Systemic fungicides oxanthin

3

Plant Disease

Plant Disease

Disease (literally *dis-ease*) implies *lack of 'comfort 'Deviation from normal functioning.*

Definitions

Disease is any morphological or physiological abnormality in a plant or any of its parts caused by continuous irritation and results in economic losses to human beings.

All these Definitions Iindicate that Disease

1. Is not a pathogen but it is caused by a pathogen.
2. It is not symptom but results in symptoms.
3. It is not a condition as the condition results from disease and is not syncnymous with it.
4. It is not an injury.
5. Cannot be infectious, it is actually the pathogen which is infectious.
6. Results from continuous irritation.
7. Malfunctioning process and this result in some suffering and therefore disease is a pathological process.

Importance of Plant Diseases

The major importance of plant diseases are :

1. Limit the kinds of plants and industries in the area.
2. Reduce the quality and quantity of plant produce.
3. May make plants poisonous to humans and animals.
4. May cause financial losses.
5. Cost of controlling plant diseases is also a direct loss to diseases.

Classical examples

S. No	Major disease outbreaks	Location/ Country	Period	Impacts
1	Late blight of potato	Ireland	1845-46	Irish famine
2	Brown spot of rice	Asia (India)	1943	Bengal famine
3	Coffee rust	Srilanka	1870s-1880s	Destroyed all coffee in South east Asia

Plants and Disease

- Plants make up the majority of the earth's living environment.
- Plants are the primary source of food for humans and animals, whether directly or indirectly.
- Plants are the only higher organisms that can convert the energy of sunlight into stored usable chemical energy in carbohydrates, proteins and fats.
- All animals including humans, depend on these plant substances for survival.
- Even the meat, milk, and eggs that we and other carnivores eat come from animals that themselves depend on plants for their food.
- Plants, however, also get sick.
- Sick plants grow and produce poorly, they exhibit various types of symptoms, and, often, parts of plants or whole plants die.
- It is not known whether diseased plants feel pain or discomfort.

Concept of Disease in Plants

Healthy Plant

- A plant is healthy or normal when it can carry out its physiological functions to the best of its genetic potential.
- Meristematic (cambium) cells of a healthy plant divide and differentiate as needed.
- Different types of specialized cells absorb water and nutrients from the soil, translocate these to all plant parts.
- Plants perform photosynthesis, transport, metabolise, and store photosynthesis products.
- Produce seed or other reproductive organs for survival and multiplication.

Diseased Plant

It is difficult to pinpoint exactly when a plant is diseased because

- It is not known whether plants feel pain or discomfort.
- Plants do not speak,or otherwise communicate with us.

- When the ability of the cells of a plant or plant part to carry out one or more of these essential functions is interfered with by either a pathogenic organism or an adverse environmental factor, the activities of the cells are disrupted, altered, or inhibited, the cells malfunction or die, and the plant becomes diseased.
- At first, the infection is localized to one or a few cells and is invisible.
- Soon, however, the reaction becomes more widespread and affected plant parts develop changes visible to the naked eye.
- The visible changes are the symptoms of the disease.
- The visible or otherwise measurable adverse changes in a plant, produced in reaction to infection by an organism or to an unfavorable environmental factor, are a measure of the amount of disease in the plant.
- Disease in plants defined *as the series of invisible and visible responses of plant cells and tissues to a pathogenic organism or environmental factor that result in adverse changes in the form, function, or integrity of the plant and may lead to partial impairment or death of plant parts or of the entire plant.*
- The kinds of cells and tissues that become affected determine the type of physiological function that will be disrupted first. For example,
- Infection of roots may cause roots to rot and make them unable to absorb water and nutrients from the soil.
- Infection of xylem vessels, as happens in vascular wilts and in some cankers, interferes with the translocation of water and minerals to the crown of the plant;.
- Infection of the foliage, as happens in leaf spots, blights, rusts, mildews, mosaics, and so on, interferes with photosynthesis.
- Infection of phloem cells in the veins of leaves and in the bark of stems and shoots, as happens in cankers and in diseases caused by viruses, mollicutes, and protozoa, interferes with the downward translocation of photosynthetic products.
- Infection of flowers and fruits interferes with reproduction.
- Although infected cells in most diseases are weakened or die, in some diseases.
- Hyperplasia: Infected cells are induced to divide much faster e.g., in crown gall.
- Hypertrophy: Enlarge a great deal more (hypertrophy) than normal cells.
- Tumors: Produce abnormal amorphous overgrowths.

Basic Function in a Plant and Interference Caused by Plant Pathogen

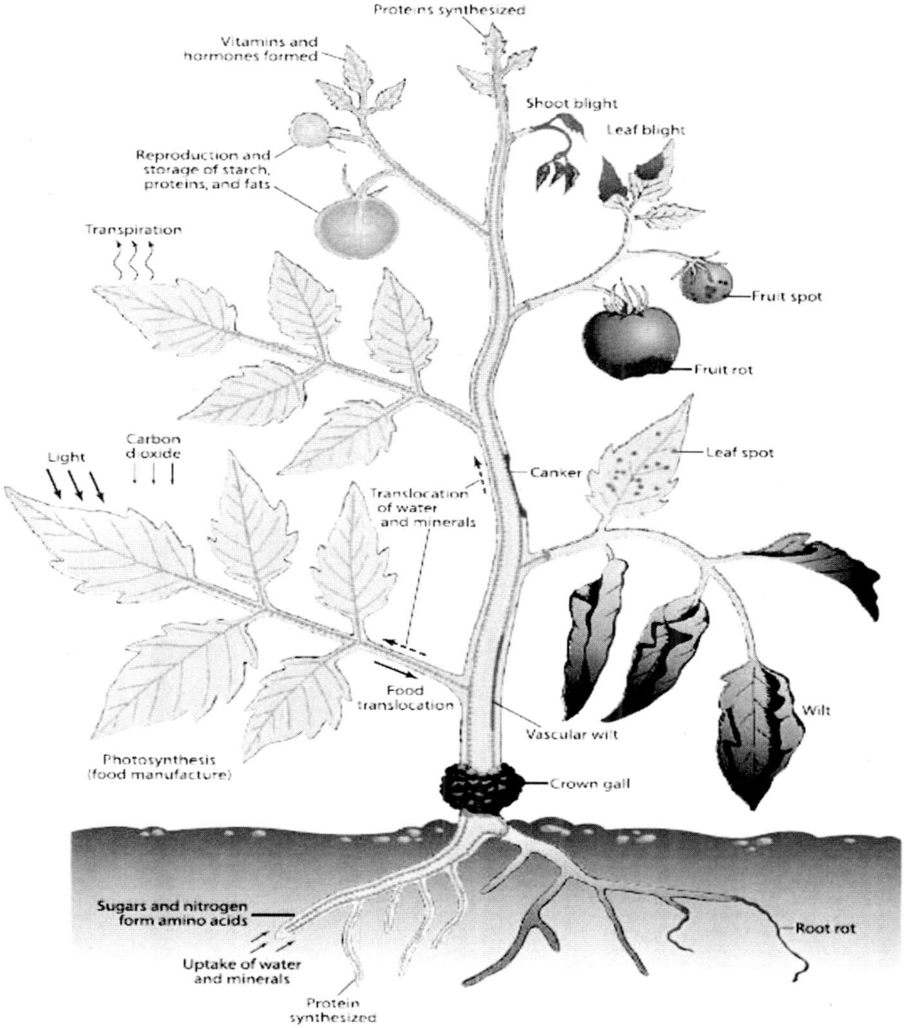

Morphology and way of multiplication of plant pathogens

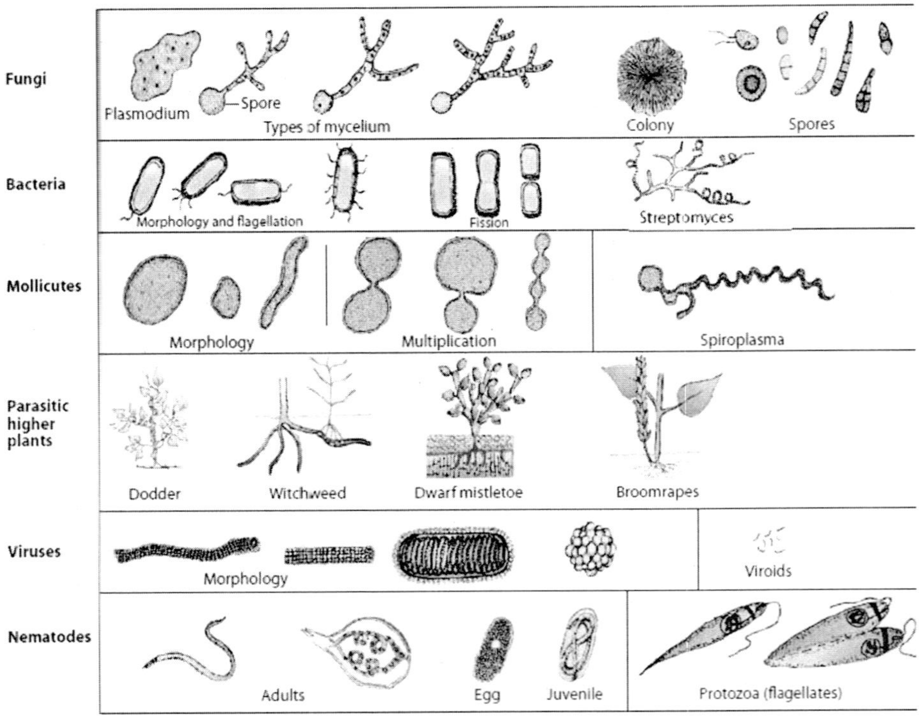

Special Information

Mistletoe

- Mistletoe recognized as the first plant pathogen.
- First pathogen for which a cultural control (by pruning affected branches) was recommended.
- Obligate hemiparasitic plant.

4

Major Epidemics and Their Social Impacts

Global & National Yield Losses Due to Pests, Pathogens and Weeds

- Yield losses caused by pests, pathogens, and weeds are responsible for 20–40% loss of global agricultural productivity.
- In India, 20-25% crop losses are reported due to pest attacks every year.
- As per Ministry of Agriculture, India is losing agricultural production worth INR 1.48 lakh crores annually due to damage from pests, weeds and plant diseases.

Major epidemics and their social impacts

Disease	Location	Year	Comments
Late blight of potato	Cool, humid climates	1845–1846	Annual epidemics, e.g., Irish famine
Brown spot of rice	Asia	1943	Epidemics, e.g., the great Bengal famine
Southern corn leaf blight	U.S.	1970	Historical interest, epidemic $1 billion lost
Downy mildew of grapes	U.S	1870s–1880s	European epidemic
Chestnut blight	U.S	1904–1940	Destroyed almost all American chestnut trees
Dutch elm disease	U.S	1918 to present	Europe. Destroying American elm trees
Coffee rust	Asia	1870s–1880s	South America. Destroyed all coffee in southeast Asia since 1970. present in South and Central America

Diseases Likely to Cause Major Epidemics and their Social impacts in Future

1. Fungal

Disease	Comments
Late blight of potato & tomato	New mating type A1 & A2 of fungus spreading worldwide
Karnal bunt of wheat	Destructive in Pakistan, India, Nepal; since the 1980s introduced into Mexico and in the 1990s into U.S
Downy mildew of corn and sorghum	Downy mildew of corn and sorghum Just spreading beyond southeast Asia
Soybean rust	Spreading from Southeast Asia and from Russia; already in Hawaii, Puerto Rico, and South America
Chrysanthemum white rust	Important in Europe, Asia, and recently in California
Sugar cane rust	-
Citrus black spot	-

2. Viral

Disease	Comments
African cassava mosaic	Destructive in Africa; threatening Asia and the Americas
Streak disease of maize (corn)	Spread throughout Africa on sugar cane, corn, wheat, etc.
Hoja blanca (white tip) of rice	Destructive in the Americas so far
Bunchy top of banana	Destructive in Asia, Australia, Egypt,
Rice tungro	Destructive in Southeast Asia
Bean golden mosaic	Caribbean basin, Central America, Florida
Tomato yellow leaf curl	East Mediterranean, Caribbean, the America

3. Others

Pathogen	Disease	Comments
Bacterial	Bacterial leaf blight of rice	Destructive in Japan and India; spreading
	Bacterial wilt of banana	Destructive in the Americas; spreading elsewhere
	Pierce's disease of grape	Deadly in southeast U.S.; spreading into California
	Citrus variegation	Destructive in Brazil; spreading
	Citrus greening	Severe in Asia; spreading
Phytoplasmal	Lethal yellowing of coconut palms	Destructive in Central America; spreading into U.S.
Viroid	Cadang-cadang disease of coconut	Killed more than 15 million trees in the Philippines to date
Nematode	Burrowing nematode	Severe on banana in many areas and citrus in Florida
	Red ring of palms	Severe in Central America and the Caribbean
	Pinewood nematode	Widespread and becoming severe in North America

5

Classification of Plant Diseases

Classification of Plant Diseases

- There are number of diseases which attacks crop plants.
- Therefore It is mandatory to classify plant disease into more manageable and more specific group based on several criteria for easy and better understanding.
- Let us discuss one by one.

Classification based on the host plant

Diseases are classified like

- Disease of apple
- Diseases of wheat
- Diseases of rose
- Diseases of coconut
- Diseases of coffee
- Diseases of cotton

Classification based on symptoms

Disease can be classified into

- Rusts
- Smuts
- Wilts
- Blights
- Cankers
- Mildews
- Rots
- Damping-off
- Die-back
- Scab etc.

Classification on the basis of type of Infection

1. **Localized disease**
   - Affecting only a part of the plant.
   - Example: Leaf spots and anthracnoses
2. **Systemic disease**
   - Affecting the entire plant.
   - Examples: Mosaic and Leaf curl disease

Classification based on their Occurrence

1. **Endemic disease**
   - When a disease is more or less constantly present in one form or other or less constantly present form year to year in a moderate to severe form, in a particular country or part of earth.
   - Ex. Wart disease of potato in Darjelling districts
2. **Epidemic or Epiphytotic diseases**
   - Term 'epidemic' means 'among the people'
   - In true sense applies to those diseases of human beings which appear very virulently among large section of the population.
   - To carry the same sense in the case of plant diseases, the term epiphytotic has been coined.
   - An epiphytotic disease is one which occurs widely but periodically.
   - It may be present constantly in the locality but assumes severe form only on occasions.
   - Ex. Irish famine due to late blight of potato
3. **Sporadic diseases**
   - Diseases which occur at very irregular intervals and locations and in relatively few instances.
   - Ex. Potato spindle tuber.
4. **Pandemic diseases**
   - When a disease is prevalent throughout the country, continent or the world.
   - Latest example corona disease.

Classification Based on the Cause

1. **Infectious or Biotic disease**
   - Diseases incited by biotic or mesobiotic agents under a set of suitable environments.
   - Example, fungal disease, bacterial disease, viral disease.

2. **Non-infectious or Abiotic disease**
   - Diseases with no biotic or mesobiotic agents associated.
   - Remain noninfectious and cannot be transmitted from one diseased plant to another healthy plant.
   - Example. Disease caused by nature-frost, rain, wind, sun, hail storm.

Classification Based on the Production of Inoculum

1. **Single cycle disease**
   - Called simple interest disease.
   - When the increase of disease is mathematically analogs to simple interest of money.
   - There is only one generation of disease in the course of one epidemic.
   - Such diseases develop from a common source of inoculums.
   - Capital i.e inoculum is constant.
   - Example. Loose smut of wheat
2. **Multiple cycle disease**
   - Compound interest disease
   - When the increase in disease is mathematically analogues to compound interest of money, the disease.
   - There are several or many generations of the pathogen in one life cycle of the crop.
   - Capital ie inoculum is increased by the amount of interest.
   - Example. Late blight of potato.

Classification on Basis of Type of Perpetuation & Spread

1. **Soil-borne diseases**
   - The causal agents perpetuate and spread through soil.
   - Example. Damping off and Fusarium wilt.
2. **Seed-borne diseases**
   - Seed or seed materials help in the perpetuation and spread of this disease.
   - The disease causing agents may be internally seed-borne or externally seed-borne
   - Example:
   - Loose smut of wheat (Example of internally seed borne)
   - Blast of rice (Example of externally seed-borne).

3. **Air-borne diseases**
   - In these type of diseases the causal agents are spread by wind (air).
   - Example.
   - Early leaf spot and late leaf spot of groundnut,
   - Rust of wheat

Classification based on sugar requirement of the pathogen

1. High sugar diseases
2. Low sugar diseases.

1. High sugar diseases

- When crops are sprayed with chemicals like DDT or maleic hydrazide which inhibit the outflow of sugars from the leaf.
- Termed these diseases as 'high sugar diseases'.
- Examples: Diseases such as rusts and powdery mildews increased.

2. Low sugar diseases

- When the crops were sprayed with chemicals like 2,4 D which increase the outflow of sugars
- Termed these diseases as 'low sugar diseases'
- Examples: disease like spots increased.

Iatrogenic diseases

- Diseases which appear on the plant while managing major diseases are called iatrogenic diseases.
- E.g. when zineb is sprayed on grapes for the management of downy mildew, the crop suffers losses from grey mold which otherwise do not cause any disease.
- This happens owing to the effects of the chemical on microclimatic and microbial flora on the leaf surface.
- The chemical may affect the structural composition or chemical composition of the sugars or the leaf exudates or the phylloplane microflora resulting in conditions favorable for the new pathogen.

6

Growth of Plant Pathogen

Growth of Plant Pathogen

Plant Pathogens

- Fungi
- Bacteria
- Nematodes
- Viruses
- Viroids
- Mollicutes
- Fastidious Bacteria
- Protozoa

These Plant pathogens grow in diverse ways

Fungi and Parasitic Higher Plants

- Fungal pathogens and parasitic higher plants normally invade and infect tissues by growing on or into them from one initial point of inoculation.
- Most of these pathogens, whether inducing a small lesion, a large infected area of the plant, continue to grow, branch out and spreads into more and more plant tissues until the spread of the infection is stopped or the plant is dead.
- Fungi causing vascular wilts often invade plants by producing and releasing spores within the vessels.
- As the spores are carried in the sap stream they invade vessels far away from the mycelium, germinate there, and produce a mycelium, which invades more vessels.

Other Plant Pathogens

- Bacteria, mollicutes, viruses, viroids, nematodes, and protozoa, do not increase much, if at all, in size with time
- Their size and shape remain relatively unchanged throughout their existence

- These pathogens invade and infect new tissues within the plant by reproducing at a rapid rate and increasing their numbers tremendously in the infected tissues.
- Progeny may then be carried passively into new cells and tissues through
- Plasmodesmata : Viruses and viroids.
- Phloem: Viruses, viroids, mollicutes, some fastidious bacteria, protozoa.
- Xylem: some bacteria, protozoa and nematodes.

Reproduction of Pathogen

Plant pathogens

- Fungi
- Bacteria
- Nematodes
- Viruses
- Viroids
- Mollicutes
- Fastidious Bacteria
- Protozoa

These plant pathogens reproduce in diverse ways

Reproduction in Fungi

- Fungi reproduce by means of spores, which may be either
- Asexual (Mitospores, i.e., Products of mitosis).
- Sexual (Meiospores, i.e. Products of meiosis).
- Majority of plant pathogenic fungi and oomycetes produce a mycelium only within the plants they infect and also produce spores on, or just below the surface of the infected area of the host.
- The spores are released outward into the environment.
- Few fungi and oomycetes produce a mycelium on the surface of their host plants, but most powdery mildew fungi produce a mycelium only on the surface of, and none within, their hosts.
- Plant pathogenic clubroot pathogen and fungi causing vascular wilts, produce spores within the host tissues.
- These spores are not released outward until the host dies and disintegrates.
- Some fungi produce spores more or less continuously whereas others produce them in successive crops.

- In either case, several hundred to thousands of spores may be produced per square centimeter of infected tissue.
- Even small specialized sporophores can produce millions of spores.
- The number of spores produced per diseased plant is often in the billions or trillions.
- The number of spores produced in an acre of heavily infected plants.
- Enough spores are released to land on every conceivable surface in the field and the surrounding areas, enough to easily inoculate with a heavy inoculum every plant in the area.

Reproduction in Bacteria

- Bacteria reproduce by fission.
- One mature individual splits into two equal, smaller individuals.
- Bacteria reproduce between or, in xylem- or phloem-inhabiting bacteria, within host cells generally inside the host plant.
- They come to the host surface only through wounds, cracks, stomata, and so on.
- Bacteria reproduce rapidly within infected tissues.
- Bacteria divide (double their numbers) every 20 to 30 minute under optimum nutritional and environmental conditions (in culture).
- Bacteria multiply just as fast in a susceptible plant as long as nutrients and space are available and the temperature is favorable.
- Millions of bacteria may be present in a single drop of infected plant sap so the number of bacteria per plant must be astronomical.

Reproduction in Viruses and Viroids

- Reproduce only inside cells and apparently do not reach or exist on the surface of the host plant.
- They are replicated by the cell, just as a page placed on a photocopying machine is replicated by the machine as long as the machine is operating and paper supplies last.
- They reproduce within living host cells, with the first new virus particles being detectable several hours after infection.
- Soon after that, however, virus particles accumulate within the infected living cell until as many as 100,000 to 10,000,000 particles may be present in a single cell.
- Viruses and viroids infect and multiply in most or all living cells of their hosts.

- It is apparent that each plant may contain innumerable individuals of these pathogens.

Reproduction in Fastidious Bacteria, Mollicutes and Protozoa-

- Mollicutes, protozoa, and fastidious bacteria reproduce only inside cells and apparently do not reach or exist on the surface of the host plant.
- Fastidious bacteria and mollicutes also reproduce by fission .
- Fastidious bacteria and mollicutes appear to reproduce more slowly than typical bacteria.
- Although they spread systemically throughout the vascular system of the plant.
- They are present in relatively few xylem or phloem vessels.
- Total number of these pathogens in infected plants is relatively small.
- This also seems to be true for protozoa.

Reproduction in Nematode

- Nematodes reproduce by means of eggs.
- Nematode females lay about 300 to 500 eggs, about half of which produce females that again lay 300 to 600 eggs each.
- Nematode species may have from two to more than a dozen generations per year, depending on the climate, the availability of hosts, and the duration of each life cycle of the particular nematode.
- If even just half of the females survived and reproduced, each generation time would increase the number of nematodes in the soil by more than a hundred fold.
- Thus, the buildup of nematode populations within a growing season and in successive seasons is often quite dramatic.

7

Koch's Postulates

Robert Koch

1. Period : 1843–1910.
2. Medical doctor and a bacteriologist.
3. First to show, in 1876, that anthrax, a disease of sheep was caused by a bacterium.
4. That he called *Bacillus anthracis*.
5. He subsequently discovered, in 1882, that tuberculosis and, in 1883, that cholera are each caused by a different bacterium.
6. This led to the general conclusion that each disease is caused by a specific microbe.
7. These experiments confirmed for the first time the germ theory of disease proposed earlier by Louis Pasteur.

Koch's Postulates

- Koch set out the four steps or criteria in 1887 that must be satisfied before a microorganism isolated from a diseased human, animal, or plant can be considered as the cause of the disease.
- These FOUR STEPS, rules, or criteria are known as "Koch's postulates."
    1. The suspected causal agent (bacterium or other microorganism) must be present in every diseased organism (e.g., a plant) examined.
    2. The suspected causal agent (bacterium, etc.) must be isolated from the diseased host organism (plant) and grown in pure culture.
    3. When a pure culture of the suspected causal agent is inoculated into a healthy susceptible host (plant), the host must reproduce the specific disease.
    4. The same causal agent must be recovered again from the experimentally inoculated and infected host, i.e., the recovered agent must have the same characteristics as the organism in step 2.

Advantage and Disadvantage of Koch's postulates

- Koch's rules are possible to implement.

- Although not always easy to carry out, with such pathogens as fungi, bacteria, parasitic higher plants, nematodes, most viruses and viroids, and the spiroplasmas.
- These organisms can be isolated and cultured, or can be purified, and they can then be introduced into the plant to see if they cause the disease.
- With the other pathogens, however, such as some viruses, phytoplasmas, fastidious phloem-inhabiting bacteria, protozoa, and even some plant pathogenic fungi that are obligate parasites of plants (such as the powdery mildew, downy mildew, and rust fungi), culture or purification of the pathogen is not yet possible and the pathogen often cannot be reintroduced into the plant to reproduce the disease.
- Despite the difficulties of carrying out Koch's postulates with some causal agents, they continue to be applied, sometimes with certain modifications, in all cases of disease.
- By attempting to carry out Koch's postulates in all newly discovered diseases, a great deal of work with potential saprophytes has been avoided.
- While, at the same time, doubt and criticism are reduced to a minimum while confidence in and use of the identification increase greatly and quickly.

Physiological Specialization

Physiological specialization

- Within the species of a pathogen there exist certain individuals that are morphologically similar but differs with respect to their physiology, biochemical characters and pathogenicity and are differentiated on the basis of their reaction on certain host genera or cultivars.

Physiologic race

- Individuals within the species of a pathogen that morphologically similar but differ with respect to their pathogenicity on particular set of host varieties.

Forma specialis (f. sp.)

- Individuals within the specie of a pathogen that morphologically similar but differ with respect to their pathogenecity on particular host genera.
- E.g. *Puccinia graminis f.sp. tritici.*

Pathotype

- A pathotype is a population of a parasite species in which all individuals have a stated pathosystem character (pathogenicity or parasitic ability) in common.

Variant

- Sometimes , one of the off springs of a race can suddenly attack a new variety that was not infected before, this individual is called a variant.

Biotype

- Progeny developed by variant having similar heredity is called a biotype or a subgroup of individuals within the species, usually characterized by the possession of single or few characters in common.

Varieties or special forms

- Some individuals of the species attack only certain species of host plants, which make up formae specialis.

8

Survival & Dispersal of Plant Pathogens

Survival & Dispersal of Plant Pathogens

- Two important aspects of any plant disease cycle are:
  - i) Ways in which the parasite survives between cropping seasons or during periods when the environment is unfavourable.
  - ii) Means by which it spreads from a diseased plant to one that is to become infected.
- If survival or spread is prevented the disease will not occur.
- Thus, a knowledge of survival and dispersal mechanisms often reveals a weak link in the disease cycle that can be utilized in managing the disease

Survival of Plant Pathogens

- Most plant pathogens can survive only by repeatedly reinfecting host plants.
- This process of repeated infection is known as an infection chain.
- If the infection chain is broken the pathogen dies out.
- The Swiss plant pathologist Ernst Gaumann (1950) in his book Principles of Plant Infection divided infection chains into two types depending on how the inoculum of the pathogen survives between leaving the infected host (the distributor host) and arriving on the host to be infected (the receiver host).
- In continuous infection chains the pathogen survives by continually infecting susceptible plants in an uninterrupted fashion.
- Discontinuous infection chains occur when the disease cycle is interrupted by an epiphytic phase, a saprophytic phase or a resting stage

Five Types of Infection Chains available to Plant Pathogens

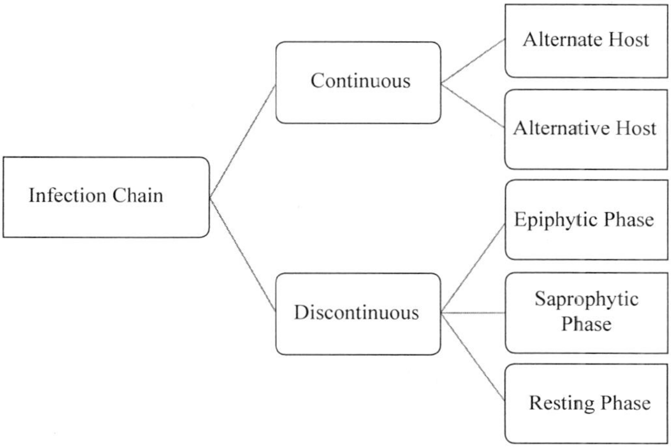

Source of Survival

1. Infected plants as reservoir of inoculum
    - Cultivated host
    - Collateral and other weed hosts.
    - Alternate hosts
2. Saprophytic survival
3. Survival through dormant / resting structures (organs)
4. Survivals with seeds
    - Fungal pathogens
    - Bacterial pathogens
    - Viruses
    - Plant parasitic nematodes
5. Survival in association with insects
6. Survival in vegetatively propagated plant parts

1. Survival of Fungi

- Fungi have evolved a great variety of mechanisms for persisting between crops.
- Fungi overwinter as mycelium in diseased tissues on perennial plants, e.g., cankers, and as spores at or near the infected surface of the plant or on the bud scales.
- Fungi affecting leaves or fruits of deciduous trees usually overwinter as mycelium or spores on fallen, infected leaves or fruits or on the bud scales.

- Fungi affecting annual plants usually survive the winter or summer as mycelium in infected plant debris, as resting or other spores and as sclerotia in infected plant debris or in the soil, and as mycelium, spores, or sclerotia in or on seeds and other propagative organs, such as tubers.

2. Soil inhabitants

- They are able to survive indefinitely as saprophytes.
- Generally unspecialized parasites that have a wide host range.
- Some plant pathogenic oomycetes (e.g., *Pythium*) and fungi (e.g., *Fusarium, Rhizoctonia*).
- They are rather specialized parasites that generally live in close association with their host.
- May survive in the soil for relatively short periods of time as hardy spores or as saprophytes

3. Soil transients

- Some fungi infect cultivated or wild perennial, as well as annual, plants and move from the perennial to the annual ones each growth season.
- Some rust and other fungi overwinter on winter crops grown in warmer climates and move from them to the same hosts grown as spring crops in colder climates.
- Some rust fungi infect alternately an annual and a perennial host, and the fungus goes from the one to the other host and overwinters in the perennial host.
- In some areas, fungi survive by continuous infection of host plants grown outdoors throughout the year, such as cabbage, or of plants grown in the greenhouse in the winter and outdoors in the summer.

Suvival of Bacteria-

- Bacteria also overwinter and oversummer in infected plants, seeds, and tubers, in infected plant debris, and, for some, in the soil.
- Bacteria survive poorly when present in small numbers and free in the soil but survive well when masses of them are embedded in the hardened, slimy polysaccharides that usually surround them.
- Some bacteria also overwinter within the bodies of their insect vectors.

Survival of Plant Pathogenic Bacteria

Plant pathogenic bacteria is group in relation to survival in soil:

1. Transient visitor
2. Resident visitor
3. Saprophytes

1. Transient visitor

- Phytopathogenic bacteria populations are developed almost exclusively in the host.
- Maximum number of generations are produced in host.
- When such bacteria reach soil, their populations decline rapidly and do not remain important source of primary inoculum.
- Being poor competitors, most phytopathogenic bacteria comes under this category.
- Example: *X. campestris pv. phaseoli, pv. malvcerum, pv. citri, pv. campestris, pv. vesicatoria* etc.

2. Resident visitor

- Also have their maximum generation in the host but their numbers gradually decline in soil.
- Slow decline would permit net increase of such bacteria from season to season in soil.
- Also called soil inhabiting bacteria
- Example; *Agrobacterium tumefaciens, Ralstonia solanacearum, Streptomyces scabies, S. ipomoea*

3. Truly saprophyte

- Very few phytopathogenic bacteria.
- Bacteria whose populations are largely produced in soil including the rhizoplane.
- Bacteria whose relationship to plant disease is ephemeral.
- Called true soil saprophytes.
- Example: Pseudomonas, Clostrodium

Viruses, Viroids, Mollicutes, Fastidious Bacteria, and Protozoa

- They survive only in living plant tissues such as the tops and roots of perennial plants, the roots of perennial plants that die to the soil line in the winter or summer, vegetative propagating organs, and the seeds of some hosts.
- A few viruses survive within their insect vectors, and some viruses and viroids may survive on contaminated tools and in infected plant debris.

Nematodes

- Nematodes usually overwinter or oversummer as eggs in the soil and as eggs or nematodes in plant roots or in plant debris.
- Some nematodes produce juvenile stages or adults that can remain dormant in seeds or on bulbs for many months or years.

Dispersal of Plant Pathogens

Definition

Distribution

- Spred of a pathogen into new geographical areas and its established there.

Dispersal

- Is what happens between take off of a spore and its deposition.
- It does not include the germination or infection of the plant.

Spread

- Implies that pathogen reaches and infects plants.
- Progressive colonization of the infected organ or plants.
- Passage of the pathogen from infected plants to others in the same field or crop area.
- Long distance spread of pathogens between plants which are widely separated, as in intercontinental spread.

Importance of Dispersal

- For continuity of the life cycle of pathogen.
- For development of the pathogen.
- For spread of diseases.

Why knowledge of pathogen dispersal is essential ?

- To find out effective control measures for diseases.
- To break the infection chain of pathogen.

Types of plant pathogen dispersal

Two types

A. Direct (Autonomous) dispersal

The pathogen is carried through

- Soil
- Seed
- Planting material

B. Indirect (Passive) dispersal

The pathogen spreading itself or by natural agencies like

- Wind
- Water

- Animals
- Insects
- Mites
- Nematodes
- Birds

A. Direct dispersal (Through soil)

Through soil

- Plant pathogens survive in soil and their dispersal takes place in two ways
- Many pathogens survive in soil as facultative parasites or facultative saprophytes.
    - Some of the fungi produce spores which are motile (Zoospores) and can move some distances in soil and infect their hosts.
    - Nematodes can move some distances in soil infect their hosts.
- The contaminated soil is transported to other places along with planting material, seedlings or implements etc. while adopting agronomic practices.

Through Seed

- Seed produced on diseased plants or on healthy plants within a diseased crop may become contaminated with or become infected by plant parasites.
- Seedborne pathogens typically survive for a long period in or on contaminated or infected seed.
- Consequently, they can be spread over vast distances in the commercial distribution of seed.
- Within a crop, infected seed usually results in the development of randomized foci of infection .
- So the pathogen must also have other means of dispersal to distribute it within the crop during the growing season.
- Inoculum of seed-borne pathogens can be transmitted in at least four different ways.
- It can
    1. Be admixed with the seed
    2. Adhere to the seed
    3. Become embedded in the seed or
    4. Invade the embryo within the seed

I. Admixed seed-borne inoculum

- This occurs when inoculum of the pathogen becomes mixed with the seed, usually during harvesting.
- In some instances inoculum is about the same size as the seed of the host plant.
- e.g. sclerotia of *Sclerotinia sclerotiorum* in sunflower seed and sclerotia (ergots) of *Claviceps purpurea* in cereal and grass seeds).
- The seeds of parasitic plants such as dodder can also become mixed with the seed of host species.
- Inoculum that becomes admixed with seed can often be physically separated from the seed by flotation (e.g. in brine), gravity methods or by sieving.

II. Adherent seed-borne inoculum

- Inocula of many fungal and bacterial pathogens are transmitted as a result of inoculum becoming attached to or adhering to the surface of seed.
- Inoculum transmitted in this way includes conidia of fungi such as Alternaria spp., Drechslera spp. and teliospores of seedling-infecting smuts (e.g. the bunt fungus, *Tilletta caries* and the flag smut fungus *Urocystis agropgri*, both of which infect wheat).
- Cells of the bacterium *Xanthomonas campestris* pv. *campestris* (the cause of black rot of brassicas) and *X. campestris* pv. *malvacearum* (the cause of bacterial blight of cotton) can be spread as adherent inoculum on seed.
- Adherent seed-borne inoculum can often be killed by treating the seed with chemicals.

III. Embedded seed-borne inoculum

- Some seed-borne pathogens actually infect the seed by penetrating the seed coat and internal tissues.
- In other words they become embedded in the seed.
- During seed dormancy the pathogen and host seed tend to co-exist.
- However, upon germination the pathogen becomes active and may kill or damage the seed before or after emergence.
- Fungi such as some species of Alternaria, Cercospora and Drechslera form a mycelium in the infected seed.
- The pycnidia of fungi such as some species of Septorta and Phoma can often be seen on infected seeds.

- Bacteria such as *Xanthomono.s campestris* pv. *malvacearum*, the cause of bacterial blight of cotton can also be transmitted as embedded inocula.
- It is often difficult to destroy inoculum that is embedded in seed because it is not destroyed by contact fungicides.

IV. Embryo-borne inoculum

- The inoculum of some pathogens becomes located in the embryo and/or the endosperm or in tissues derived from the embryo sac.
- Inoculum of floret infecting smuts such as *Ustilago tritici* and some bacteria such as *Corynebacterium michganense* pv. *michiganense* can be transmitted in the embryo of seeds.
- About one fifth of the known plant viruses are seed-borne.
- Most cases of seed transmission occur as a result of infection of the embryo and only occasionally has carryover of virus on the testa been shown to be responsible for the infection of seedlings.
- Infected seed can result either from infection of the ovule from the maternal parent or by pollination of the flower of a healthy plant with infected pollen.

Through vegetative planting materials

- Dispersal of large number of plant pathogens takes place through the vegetative parts of the plants used as planting materials, such as tubers, cuttings, runners, rhizomes, grafts, etc.
- Thus virus diseases are mainly severe in vegetative propagated horticultural crops.
- Over 40% of the bacterial plant pathogens are transmitted on vegetatively propagated materials.
- Among the important examples are
- *X. compestris* pv.*citri* (citrus canker)
- *A. tumifaciens* (crown gall).

Autonomous dispersal of plant pathogens

Autonomous dispersal	**Examples**
Through seed	
Seed contamination	*Anguina tritici* (seed gall nematode), *Tolypo-sporium penicillariae* (Smut of bajra), *Claviceps, Cuscuta*
Internally seed borne	Loose smut of wheat *(Ustilago tritici)*

Externally seed borne	Covered smut of barley (*Ustilago hordei*), Stinging smut of wheat (*Tiilletia caries and Tilletia foetida).* Black arm of cotton *(Xanthomonas campestris* pv. *malvacearum*
Through vegetative planting materials	
Seed tubers	Late blight of potato (*Phytophthora infestans*)
Sugarcane Sets/cutting	Red rot of sugarcane(*Colletotrichum falcatum*)
Propagative materials	Citrus *Tristeza* and greening
Rhizome	Rhizome rot of ginger

B. Indirect dispersal

Wind dispersal

- Extensive and severe epidemics of plant diseases are mostly the results of wind dispersal of the pathogens.
- Fungal spores produced externally on host surfaces are most easily carried by wind currents.
- Wind dispersal involves four stages relating to the spores *viz.*
  - Production of countless spores.
  - Their liberation in the wind currents.
  - Dispersal along with the wind.
  - And deposition on new susceptible host surfaces where they cause infection under favourable climatic conditions.
- This is the most dangerous mode of transmission of plant pathogenic fungi.
- Examples : Pathogen causing powdery and downy mildews, leaf spots, blasts, blights and rust diseases.

Water dispersal

- Comparatively less important as compared to the wind dispersal.
- Splashing rain drops generally transmit the foliar diseases from
  - Leaf to leaf
  - Shoot to shoot
  - Plant to plant in case of closely spaced crops.
- Plant pathogens requiring high humidity conditions like the fungi causing downy mildew diseases or bacteria causing canker of citrus are well adapted to this kind of short distance water dispersal.
- Certain soil inhabiting pathogenic fungi and bacteria causing root and collar rots, wilts, foot, rots, etc are likely to be transmitted to much

longer distances through the agencies like irrigation water, streams and rivers, etc.

- It is also an important agency in transmission of seeds of higher flowering parasites like dodder and striga.

Passive (Indirect) Dispersal of Plant Pathogens

Passive dispersal		**Examples**
Through water	Water splash	Bacterial leaf blight of rice *(Xanthomonas campestris pv. oryzaecola),* Bacterial streak disease of paddy *(Xanthomonas oryzecola)* Coffee rust *(Hemileia vastatrix)*
	Water flow dispersal	Club root of cabbage (*Plasmodiophora brassicae)* Late blight of potato (*Phytophthora infstans*), Red rot of sugarcane (*Colletotrichum falcatum*)
Through Air		Scab of apple (*Venturia inaequalis*), Black stem rust of wheat (*Puccinnia graminia tritici*), Late blight of potato (*Phytophthora infestans*)
Through insects	Bees and wasp	Fire blight of apple (*Erwinia amylovora*)
	Beetle (*Scolytus*)	Dutch elm disease (*Ceratocystis elmi*)
	Leaf minor	Citrus canker (*Xanthomonas citri*)
Through nematodes	*Anguina tritici*	Yellow ear rot of wheat (*Corynebacterium tritici*)
	Aphlenchoidea	Leafy gall (*Corynebacterium tritici)*
	Xiphinema index	Grapevine fan leaf virus
Through Fungi	*Synchitrium endobioticum*	Potato virus X, Potato mop top virus
	Olpidium brassicae	Tobacco necrosis virus, Letuce big vein virus
Through birds		Loranthus (*Dendrophthoe*), Mistletoe (*Viscum*), Chestnut blight (*Endothea parasitica*)
Through man		Wart of potato (*Synchitrium endobioticum*), Flag smut of wheat (*Urocystis tritici*)

9

Disease Triangle and Tetrahedron

Disease Triangle and Tetrahedron

- Disease triangle is one of the first concepts encountered by students in plant pathology.
- Interactions of the three component namely (1) susceptible host (2) virulent pathogen and (3) favorable environments have been visualized as a triangle, generally referred to as the "DISEASE TRIANGLE".
- Some plant pathologists have elaborated on the Disease Triangle by adding one or more parameters. Suggested additional parameters may be human activity, vectors and time.
- A three dimensional disease triangle may result in four dimensional DISEASE PYRAMID or DISEASE TETRAHEDRON after addition of a single parameter.
- Development of epidemic requires interaction of a highly virulent pathogen and a susceptible host in an environment that favours the development of disease.
- An understanding of these factors and their interactions for a particular disease in a particular locality allows prediction of disease outbreaks and intervention to reduce the amount of disease.

Disease Triangle

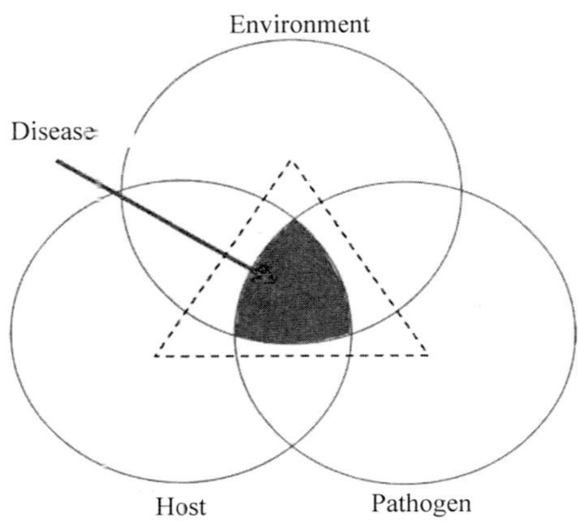

Factors Affecting Disease Development

There are three major factors that affect disease development

1. Pathogen factors
2. Host factors
3. Environmental factors

1. Pathogen factors

i. Presence of pathogen
ii. Pathogenicity (virulence and aggressiveness)
iii. Adaptability
iv. Dispersal efficiency
v. Survival efficiency
vi. Reproductive fitness

2. Host factors

i. Susceptibility
ii. Growth stage and form
iii. Population density and structure
iv. General health

3. Environmental factors

i. Temperature
ii. Rainfall
iii. Dew
iv. Leaf wetness period
v. Soil temperature
vi. Soil moisture
vii. Soil fertility
viii. Air pollution

1. Pathogen Factors

1. Presence of pathogen

- Main factor which determines whether or no disease occurs.
- In the out break of late blight of potato in north India, the presence of absence of the pathogen is the overriding factor that determines whether or not disease occurs.

2. **Pathogenicity**
   - Determines amount of disease that develops the prevalent population of the pathogen.
   - Pathogenicity comprises both the virulence of the pathogen and its aggresivenenss.
   - It also includes aspects of dispersal and survival fitness.
3. **Adaptibility**
   - Adaptability of pathogen is very important in determining its ability to overcome resistance in newly released varieties .
   - Governed by the genetic flexibility of the pathogen population and its reproductive efficiency.
   - *Puccinia graminis* has high reproductive efficiency and so is likely to adapt very fast.
   - Where as O. *theobromae* on cocoa has a much lower reproductive capacity and so likely to adapt more slowly.
4. **Dispersal efficiency**
   - Determines ability of a pathogen to cause destructive disease.
   - Urediniospores of rust can be blown long distances in a few days, allowing epidemics to develop rapidly over very large areas.
   - Soil borne pathogens have limited dispersal ability and ted to cause localized disease outbreaks.
5. **Survival efficiency**
   - Ability of a pathogen to cause disease in successive seasons depends on its ability to survive.
   - *Plasmodiophora brassicae*, the cause of club roots of crucifers, form spores that enable the pathogens to survive in the soil for many years, thus reducing the effectiveness of disease control using crop rotation.
   - The sclerotia of *Sclerotium rolfsii* and Sclerotinia spp. are also particularly effective in enabling these pathogens to survive through the summer and winter, resulting in larger quantities of inoculums to initiate epidemics in the new growing season.

2. Host Factors

1. **Susceptibility**
   - Occurrence of susceptible host population to the particular pathogen is the main factor affecting disease development.
   - Most important means of disease management is to plant varieties that are not susceptible to the prevailing population of the pathogen of concern.

2. **Growth stage and form**
   - Greatly influence the occurrence of disease.
   - Growth stage also determines the degree of closure of the canopy, which in turn affect the microclimate within the canopy.
   - Some diseases such as damping off are more common in seedlings, while others are characteristic of mature plants.
3. **Population density and structure**
   - Characteristics of population structure and density in the host impcts development of disease in the population.
   - Many plant diseases are much more common in dense than in sparse plantings.
   - Modern intensive agriculture provide and ideal situation for development of plant disease because variety.
   - Spectecular disease epidemics have occurred when susceptible host varieties were planted over large areas or when a new pathogen was introduced.
   - Such a situation occurred with the spread of late blight of potato to Ireland in the 1840s and with the spread of coffee rust to Ceylon in the 1870s.
   - Both pathogens encountered extensive monocultures of their respective hosts, potato and coffee and caused destructive epidemics.
4. **General health**
   - Prior health of the host often determining the occurrence of disease.
   - Necrotrophic pathogens are often more damaging on poorly growing than on vigorous hosts while opposite is the cause of biotrophs.

3. Environmental factors

Divided into micro, meso and macro-environemnt.

Microenvironment

1. Space in which the epidemiologic processes at cell and organ level occur.
2. Largest part of the infection cycle, from *deposition to takeoff*.

Mesoenvironment

1. Formed by the crop.
2. Influences the plants and the epidemiologic processes in the crop.

Macroenvironment

1. Air layer from the crop surface to the troposphere.
2. Some epidemiologic process occur in the macroenvironment.

Micro-environment

Phyllosphere	**Rhizosphere**
• Microenvironment of the leaves.	• A *microenvironment of the root.*
• Laminar air layer of up to 1mm in thickness surrounding the leaf.	• Hiltner (1904)
• A *three dimensional space.*	
Phylloplane	**Rhizoplane**
• *Two dimensional space*	*A two dimensional space*

Temperature

Temperature affects

Latent period	***Generation time***	***Infectious period***
The time between infection and the first appearance of disease symptoms.	The time between infection and sporulation.	The time during which the pathogen continues producing propagules.

Higher temperatures

- Disease cycle is speed up.
- Epidemics develop faster.

Under cooler conditions

- Epidemics progress is usually slower.
- Disease incidence and severity may not reach the threshold level.

Moisture

1. Influence disease outbreak caused by fungi, bacteria and nematode.
2. Cover rainfall, relative humidity, dew and leaf wetness.
3. An important effect on dispersal of pathogens.
4. Leaf wetness is a more accurate predictor of disease than rainfall and dew period.
5. Free water or raindrops facilitates the liberation and dispersal of many fungi and nearly all bacteria.
6. Combination of leaf wetness and ambient temperature is critical in determining the proporation of pathogen propagules that infect the host.

Unit 2: Laboratory and Analytical Techniques

1

Preparation and Sterilization of Common Media

Microbiological Culture Media

- Many different culture media exist, and some are better suited to particular microorganisms than others.
- No single culture medium is suitable for all microbes.
- In fact, many microbes have not yet been grown on culture media.
- Microbiological culture media provide essential substances including inorganic salts, a carbon source for energy (often a sugar, such as sucrose or glucose) and various other organic molecules necessary for microbial growth, such as amino acids and vitamins.
- Many fungi and bacteria are capable of synthesizing all the cellular metabolites required for growth.
- Therefore, can grow on media that simply provide a carbon and a nitrogen source.
- Media can be made in either liquid (broth) or solid form.
- Broth can be solidified by adding agar (usually at 2%).
- Agar is a non-utilized carbohydrate from marine algae.
- Much like gelatin, agar added to a liquid medium is usually dissolved by heating to boiling.
- Media, once prepared, is sterilized, cooled to about 121.6°C for 15 minutes, and dispensed into sterilized Petri dishes (covered plates).
- Molten agar solidifies at about 40-45°C.
- A balanced chemical composition substance or solution used for the culture of micro-organism is called MEDIUM (pl. media).
- Based on the materials used, a medium may be Natural, or Semi-Synthetic, or Synthetic.
- Natural media are more favourable to growth and sporulation than synthetic ones.

- *When preparing a media, it is helpful to remember following points.*
  - Fungi usually grow best in a rich carbohydrate medium.
  - Fungi usually prefer a slightly acid reaction, pH 6-6.5, while Bacteria often prefer pH 7.
  - Media should not be over sterilized.
  - Agar is slow to dissolve thoroughly (1-2 h boiling).
  - Agar does not solidify satisfactorily in very acid or alkaline solutions.
  - Peptone may generally be omitted from fungus media.
  - Tap water is often preferred to distilled water as it contains useful trace elements.

Types of Culture Media

Complex media

- Contain undefined compounds.
- Example; Potato Dextrose Agar (PDA) and Nutrient Agar .
- We do not know the exact chemical constituents of either the potatoes or the nutrient broth.
- PDA is a general purpose media for fungi, but bacteria will also grow on it.

Selective media

- Used to isolate a target organism, and no other organisms.
- Media are designed to enhance growth of the target organism while inhibiting growth of non-target organisms (contaminants).
- For example, lactic acid can be added to PDA to make acidified PDA (APDA).
- The lower pH of APDA (compared to PDA) promotes growth of fungi in preference to bacteria.

Differential media

- Differentiates between the target organism and all others that may grow on that particular medium.
- Medium may contain a dye that only the target organism can take up or activate.
- Colonies of the target organism will be stained by the dye, while other organisms will not be stained.
- MacConkey's agar contains a dye which stains colonies of lactose fermenting bacteria pink or red, while other bacterial colonies remain white.

- Medium may enhance particular characteristics of the target organism.
- King's B medium enhances the fluorescence of some *Pseudomonas* spp.

Synthetic media (or defined media)

- Synthetic media have a defined set of chemical constituents.
- Exact compounds and their exact amounts are known.
- We will not use any synthetic media.

Sterilization of Media and Tools

1. Sterilisation

- Complete destruction of all forms of life present in/ on the glasswares and other materials like medium employed for cultivation of the microorganisms in known as sterilization.
- Sterlization is achieved by exposing materials to lethal agents.
- Lethal agents which may be chemical, physical, or ionic in nature or, in the case of liquid, physical elimination of cells from the medium.
- Methods commonly used for sterilization are heat, gas, and in the case of liquids, ultra filtration.

Red heat

- Inoculating needles etc. can be sterilised by heating on flame to redness.

2. *Dry heat*

- Used for the sterilization of glassware, metal instruments, certain plastics, and heal stable compounds.
- Glassware can be sterilized in a hot air oven equipped with a thermostat.
- Dry heat sterilization using paper should be done at low temperature and for longer times than if metal containers are used.
- Dry heat requires higher temperatures for longer duration than moist heat for sterilization because heat conduction by the former is slow and has less penetrating power than the latter.

3. *Moist heat*

- Moist heat is usually provided by saturated steam under pressure in an autoclave or pressure cooker.
- Under pressure, moist media can be sterilized by heating at 15 p.si. for15 min to 20 min. in an autoclave or domestic pressure cooker.
- Moist heat has advantage over dry heat as conduction of heat is rapid and has greater penetrating power.
- Further, the temperature required for sterilization is lower and the duration of exposure is shorter.
- For most purposes 15 min. at 121°C or 30 min. at 115°C are recommended.

4. *Surface sterilization*

- Chromium plated or stainless steel instruments such as scalpels, forceps, scissors etc. should be dipped in alcohol and flamed.
- Hypochlorite is widely used as surface sterilization agents.
- Ultraviolet (UV) and near ultraviolet (NUV) wavelengths of less than 340 nm usually induce sporulation, but excessive dosages may inhibit it.
- Red light rarely induces sporulation.

Sterilization Methods for Tools

Flame Sterilization –

- Hold the end of the tool in the tip of the lower cone of the flame of a Bunsen burner until the tool is red-hot.
- Remove the tool and allow it to cool for about 10 seconds or by touching it to sterile agar several times.
- Cooling is important as you don't want to fry your culture!

Alcohol Sterilization

- Dip the end of the tool in alcohol (95% Ethyl alcohal) and then pass it briefly through a flame.
- You will see the alcohol burn off from the tool.
- Wait a few seconds to allow the tool to cool before you use it on your culture.
- Note that the alcohol, not the flame, sterilizes the tool, so it is not necessary to hold the tool in the flame.

Working with "Sterile Technique"

Laminar Flow Hood

- The laminar flow hood, in which air is forced through many layers of fine filters to exclude airborne contaminants, provides a sterile environment in which to pour culture media, prepare cultures, and transfer cultures to fresh media.
- Also, if one wishes to use media which have been stored in a refrigerator, the sterile hood provides a space to dry off the water which condenses on the plates as they warm to room temperature.

Working at the bench

- Sometimes, culture work must be done in an open room, as in this laboratory.
- Under these circumstances additional precautions are necessary.

- Before you do any culture work in this laboratory, wipe your bench top working surface with a 10% bleach solution.
- Take caution that you do not bleach your clothes or get bleach into your eyes.

Standard Techniques Used for Culturing Microorganisms -

1. Three-way streak

- Used to produce a pure culture of bacteria on solid media.

2. Construction of a moist chamber

- Used to induce sporulation of fungi

Preparation of Media

Preparation of Potato dextrose agar (PDA) media

- Scrub the potato clean, do not peal and cut into 12mm cubes.
- Weigh out 200 g, rinse rapidly in running water, place in water and boil until soft.
- Strain mesh, and squeeze as much of the pulp as possible through cheese cloth.
- Add 20 g dextrose and 20g agar and stir till dissolved.
- Make up to one litre and sterilize at 15 p.s.i. for 20 min.

Sterilization of Media

Autoclaving

- An autoclave uses superheated steam to kill microorganisms.
- Autoclave reaches temperatures of 121°C at 15 psi pressure, and these conditions are maintained for 15-20 minutes.
- It is the elevated temperature and not the higher pressure that kills the microbes.
- This is sufficient to kill even the endospores formed by some bacteria, which can survive boiling.
- Culture media, glassware and tools (e.g., transfer loops, razor blades) are sterilized by autoclaving.

Solid Nutrient Media in Petri Dishes

- Once the medium has been selected, the ingredients are measured in to 1 flasks in an amount to yield 500 ml with distilled water.
- The flasks should be no more than half full to prevent boiling over in the autoclave.
- Autoclaving will dissolve and disperse the ingredients.

- After complete sterilization of media in an autoclave and Petri dishes in an oven, the medium is poured into glass/ disposable plastic Petri dishes to a depth of about 5 mm.
- *All the following activities are followed under laminar air flow.*
    - Wipe the table top with rectified sprit.
    - Put on U.V. lamp on for 5 min.
    - Wash hands with rectified spirit and air dry.
    - Light the burners.
    - Arrange sterile Petri dishes near the burner.
    - Hold the flask with warm PDA and remove the cotton plug near the flame.
    - With left hand, raise one side of the Petri plate cover just enough to insert the mouth of the flask and pour 20 ml of medium.
    - Replace the cover immediately and gently rotate the dish so as to spread the medium evenly.
    - Hold open mouth of the flask near the flame and retain the plug in the left hand. Similarly pour some Petri dishes with medium.
    - Allow the plates to cool sufficiently for the medium to harden, before using.

Solid Nutrient Media in Test Tube Slants

- An agar slant is similar to that for Petri plates.
- Major difference is that the medium is heated to dissolve the ingredients and then poured about one third fill into test tubes.
- Test tubes are then plugged with cotton and autoclaved.
- Culture tubes with PDA are placed in a slanting position on wooden blocks to set the agar in a slanting position.
- Slanted medium allows more surface area on which inoculated microorganisms can grow.

Standard Techniques Used for Culturing Microorganisms

1. Three-way streak

Used to produce a pure culture of bacteria on solid media

2. Construction of a moist chamber

Used to induce sporulation of fungi

2

Methods of Isolation, Identification and Preservation of Pathogens

Growing Microorganisms in Culture

- Microorganisms do not exist in isolation from each other in nature.
- To study a particular microorganism, we must first separate it and produce a pure culture (containing no other microorganisms).
- There are several ways to isolate your organism of interest, some of which are better suited to certain groups of microbes.
- All these methods require the use of microbiological culture media and sterile techniques.

Methods of Isolation of Pathogens and Their Identification

- Separation of a pathogen from its host and its culture on a nutrient medium is called as ISOLATION.
- Diagnosis of a particular disease and isolation of the pathogen is required for identification of the pathogenic agent.
- Viruses, MLO and many fungi, particularly rust and powdery mildews, will not grow in common culture media.

A. Isolation of Fungal Pathogen from Infected Leaf

Following steps are used for isolation of fungal parasites from infected leaf.

- Collect infected leaf and cut 5-10 mm^2 sections from the margin of the lesions, in the transition tissues between healthy and dead tissues.
- Arrange four sterile Petri dishes in a row near the flame.
- Fill mercuric chloride solution in the first plate.
- Pour aseptically sterile water in rest of the dishes.
- Small sections of tissue are placed in mercuric chloride solution for 1 to 2 minutes.
- Using a sterile forceps, wash the sections in two changes of sterile water.
- Thereafter, the sections should be blotted dry on sterile paper

- After blotting dry, place 2-4 surface sterilized bits on PDA plate. Incubate at 25°C for growth.
- After 72 hrs, examine the fungal growth association with the bits and aseptically transfer to PDA slants to get a pure culture.

B. Isolation of Bacterial Pathogen from Infected Leaf

Following steps are followed for isolation from bacterial infected leaves.

- Cut a small portion of the bacterial infected tissue from the advancing lesion.
- Wash in one or two changes of sterile water and blot on sterile paper towel.
- The sections are macerated with a sterile glass rod in a small quantity of sterile water in sterilized Petri plates
- Streak a loopful of the suspension on nutrient agar plates.
- The streaking may be done by gently stroking the needle loop on the surface of the medium in a zig-zag manner.
- Incubate the Petri plates at 25°C . After 48 hrs, selected isolated colonies are transferred to nutrient agar slants and grown for future use.

Serial dilutions

- Gently touch the loop of the sterile inoculating needle to the surface of a 48 hr colony of the bacterium and shake vigorously in a sterile water column.
- Make 3 or 4 serial dilution from SI, no. 1 by transferring 1ml suspension from one tube to the next.
- Place 0.5 ml of each dilution into separate Petri dishes.
- Add melted but cool agar stir gently and allow to solidify.
- Incubate the plates at 25°C.
- After 48 hrs, examine for appearance of the single colonies at one or more of the plates.
- Pick well isolated single colonies resembling the parent cultures, transfer to nutrient agar slants and allow them to grow.
- Compare the colony characters and microscope details of the bacterium with the authentic description.

C. Isolation of Soil Fungi

Soil dilution plates

- Soil fungi have been most frequently isolated by the dilution technique as described for the purification of fungus cultures.

Soil plates

- Small soil sample (0.005-0.015gm) suspensions are dispersed across the bottom of sterile Petri dishes and melted, cool agar is add.
- Soil particles are distributed throughout the medium by rotating the dish.

Hyphal isolation method

- This method relies upon the fact that when a soil suspension is prepared many of the fungal hyphae remain with the heavier soil the suspension; a 50μm pore diam. sieve may also be used.
- Residue is examined microscopically for the presence of fungal hyphae which are removed with sterile forceps to a sterile Petri dish (usually 20-25 per dish).
- Melted, cooled agar is poured over the hyphal clumps and the plate rotated to disperse the hyphe.
- Plates are examined daily and hyphal tips isolated.

D. Isolation from Aerial parts

- After thorough washing, thin slices are cut from the advancing edge of the rot, usually marked by discoloration.
- These are floated in shallow water and treated as for root pieces.
- *These methods can also be used for separating mixed cultures; for example,*
    - A species with motile spores can be separated from one with non-motile spores by picking up the zoospores on bait,
    - or two zoosporic species can be separated if they produce,
    - or can be induced to produce (e.g. by growth at different temperature) sporangia and zoospores at different times.

E. Isolation from Roots

- Roots are first washed in running water over the night if much bacterial contamination is anticipated, and then the whole system (if small) or representative parts (if large) are submerged in shallow pond water mixture in 9 cm petri dishes.
- Surface sterilization is not recommended unless there is very heavy contamination.
- After 24 h and on successive days the dishes are searched under binocular microscope for hyphae and sporangia, which grow out readily into the water.

- Pieces of root bearing these are picked out, put in fresh water, baited usually with small pieces of grass blade (boiled for 10 min.) or with small seeds (cold sterilized with propylene oxide).
- If the root pieces appear to contain only one readily growing species, small sections (2 or 3 mm.) plated out will usually yield monocultures.
- Otherwise the bait can be renewed and plated when it appears to bear a single strain

Preservations of Pathogens

- Following methods can be used for preservation of fungi and bacteria.
- Fungal and bacterial cultures can be preserved in sterile mineral oil.
- The oil can be sterilized by autoclaving for 45 min at 121 °C. Culture slants are then covered to a depth of 1 cm above the upper most edge with sterilized mineral oil. Culture remain viable for upto two years or more at 10°C.
- Lyophilization has become one of the most popular methods of preservation. This process involves freezing the spores or cells in a colloid such as skim milk or in a sugar solution . The sample is then dried from the frozen state and sealed under vacuum in ampules.
- Petri dish cultures, diseased specimens and fruiting bodies are frozen and then placed directly in a freeze –drying apparatus. When dry, the specimens can be placed in boxes for future examination.
- The viable spores are added to 2-3 ml of sterile nonfat dry milk and then distributing about 1 ml of the suspension over frozen half filled tubes of silica gel (6-12 mesh, grade 40). Storage should be in a refrigerator. New cultures can be started by shaking a few silica gel particles from a test tube onto sterile medium. This technique has proven useful in maintaining stock cultures of fungi in viable but inert condition for several years.
- Large fruiting structures like mushrooms and conks can be preserved through air drying. Conks do not lose their shape, but mushrooms generally collapse and become distorted..

3

Methods of Inoculation

Methods

1. Soil inoculation
2. Seed inoculation
3. Stem inoculation
4. Leaf inoculation or Seedling inoculation

Soil inoculation

- For proving pathogenicity of soil- borne fungus, the fungus is first brought into pure culture using sand- maize meal or any other suitable medium.
- Fungus along with the medium is mixed with the sterilized soil in pots.
- Surface sterilized seeds or seedlings are then sown/ planted after about a week of establishment of inoculum in the soil.
- Alternatively, liquid fungal culture or spore suspension may be put directly over the exposed portion of the roots and then covered with soil in a pot.
- The symptoms thus, may develop after the necessary incubation period depending on the host pathogen interaction in a particular set of environment.

Seed Inoculation

1. Seedborne micro-organism can be proved to be pathogenic when the fungus is brought into culture from the contaminated or infected seeds.
2. These cultures are mass multiplied on a suitable substrate.
3. Then inoculated on surface sterilized seed by immersing in the spore/ mycelial suspension of the pathogen (test fungus) for the few minutes.
4. Inoculum may be loaded on seed surface with the help of vacuum pump.
5. This helps to adhere the inoculum more firmly on seed surface.
6. This technique is more useful to inoculate the fungal inoculum predominately in the form of dry spores of Aspergillus, Penicillum, and Trichoderma sp etc.

7. Such seeds are then sown in sterilized soil in pots under controlled conditions.
8. Appropriate negative control treatments should be kept for comparison for proving pathogenicity of the fungus.
9. Some special techniques are necessary in the case of **loose smut of wheat,**
   - Whole inflorescence at the susceptible age is inserted into a tube with spore suspension under a partial vacuum to infect the embryos.
   - Such a technique could also be useful for fungus like Claviceps species causing ergot disease of bajra.

Leaf inoculation or Seedling Inoculation

1. For plant pathogenic fungi, the seedling inoculation test can be carried out very successfully under artificial leaf inoculation conditions.
2. For this purpose, a unit of five plant seedlings per pot are raised in five pots of 15 cm diameter using field soil with proper agronomic management.
3. When the seedlings become 25-30 days of age, sufficient nitrogen through urea may be given in pots.
4. Within a week the seedlings begin growing vigorously with leaves turning dark green.
5. Plants, at 3 to 5 leaf stage become ready for inoculation particularly for pathogens like Alternaria species.
6. For inoculation, spore suspension is prepared from freshly developed conidial growth using sterile distilled water and strained through muslin cloth or centrifuged at 3000 rpm.
7. The spore suspension is then adjusted to a definite concentration ie. 10-20 spores per low power (100-150 X) microscopic field in a standard size droplet.
8. This is equivalent to concentration of 50000 - 1000000 spores per ml.
9. The plants are actually inoculated by spraying the spores onto the leaves using a fine atomizer at a dosage of 100 ml per 100 plants and inoculations are carried out in the evening after the sunset.
10. Use plants inoculated with a virulent, known culture as a positive treatment and plants sprayed with only distilled sterile water as a negative control.
11. The pots should be properly arranged to a compact block with respect to positive treatment and negative control plants with proper labelling.

12. Both sets of pots i.e. positive treatment and negative control plants must be kept in moist chambers with relative humidity at 90% or at saturation point at maximum temperature of 20-25°C and the minimum of 6 -10°C for 48 -72 hrs.
13. The plants are then removed from the moist chambers and transferred to green house at 20-25°C with a light duration of 12-16 hours for symptom development.

Stem Inoculation

1. Technique can be more appropriate for stem rot diseases.
2. For this purpose, place a droplet of fungal inoculum (spore suspension) containing (105-106 propagules/ml) on to the stem surface, usually on the leaf axis.
3. Stab through the droplet into the vascular tissue with a sterile needle (or tooth pick) taking care not to pierce the opposite side of the stem.
4. Alternatively, the eye of a large sewing needle which is charged with the propagule (fungal spores or mycelium) may be stabbed through the stem into the vascular tissue.
5. Cover the plant with a polythene bag for 24 hours to prevent the droplet from drying out.
6. Incubate at temperature appropriate to the disease development and observe the plants for up to several weeks or even months (in the case of fruit tree) for symptom development as in case of discoloration, necrosis etc.
7. In a similar way as described above, inoculation of cuttings and root inoculation can be done by dipping these plant portions in inoculum suspension (105-106 propagules /ml) for 5-10 minutes.
8. Such inoculated plant parts could then be potted in a suitable sterile growth medium keeping appropriate check treatments.

4

Measurement of Plant Disease

Importance

1. To direct activities of research and extension to those that are most harmful to economy.
2. To decide which of the control measures to use, one that is costly but efficient, the less expensive and less efficacious etc.
3. To obtain quantitative data in research to compare susceptibility of varieties, fungicides, environment and the like.
4. To assist in minimizing the use of dangerous chemicals in agriculture and perhaps one of.
5. To assist in the development of limited investigational resources to the best advantage i.e. in the disease resistance or in the provision of forecasting
6. To help better understanding of the disease itself.
7. To provide excellent training for students.

Measuring Disease

While measuring disease, one is interested in measuring

- **Disease incidence**; Proportion of a plant community that is diseased.
- **Disease severity**; Proportion of plant area that is affected.
- **Yield loss**: Proportion of the yield that the grower will not be able to harvest because the disease destroyed it directly or prevented the plants from producing it.

Methods of Measuring Plant Diseases

1. Recording the percentage of diseased plants, parts or organs

- This method is most useful and reliable in diseases like damping-off, seedling blight, cereal smuts, wilts and certain viral diseases.

2. Descriptive scale

- This method of recording disease includes the rating of disease severity on numerical and/or subjective scales.

- Such scales are widely used but from one disease, worker, location and season to another.

3. Logarithmic scales

- This scale reads the diseased tissues in logarithmic units below 50% and healthy tissues in the same unit above 50%.

4. Standard diagrams and field keys

- For rapid visual assessment of a disease on whole plant or plot or field, Keys are devised suitable for use to all workers.
- This includes the use of photographs, drawings and preserved specimens etc.

Computation of Disease Index –

$$\text{Disease index} = \frac{\text{Frequency} \times \text{Mean of rating category}}{\text{Total number of observations}}$$

$$\text{Disease index} = \frac{\text{Sum of all disease rating}}{\text{Total number of observation (sample)} \times \text{Maximal disease rating grade}} \times 100$$

$$\text{Disease index} = \frac{\text{Field class rating X Number of hectare in a class}}{\text{Total number of hectare}}$$

Steps for assessment of the effect of disease on crop yield –

Involves five steps

- Developing a descriptive growth stage key for the particular crop species.
- Developing methods to assess the incidence and severity of disease.
- Developing statistically sound methods of sampling crop populations for assessment of the amount of disease.
- Estimating the negative impact of particular levels of the disease on crop yield and quality, and
- Evaluating the economic benefit from various methods available for reducing the amount of disease.

Assessments

Crop Growth & Development

- Initial step to quantify the effect of disease is to develop a key that describes the growth and development of healthy plants during the growing season.
- It should describe development, either from sowing to harvest, in the case of annual plants, or from season to season in perennial plants.

- Details drawings or photographs, showing characteristics of the various stages of development, including leaf formation, flowering, fruiting and senescence are needed.
- Standardized growth keys have been developed for a number of crop plants, enabling comparison between different countries and different conditions.

Disease Incidence & Severity

- Whether, disease incidence or disease severity, or both, that are measured, depends on the nature of the disease.
- An "all or nothing" disease, for example, that inevitably kills any plant it infects, could be measured just by counting the number of plants that infected (disease incidence).
- However, in the case of a disease that causes varying degrees of damage to plants throughout the crop, a more complex measurement is needed, that assesses disease severity.
- Disease incidence for biotrophic pathogens can be measured by counting the number of plants, leaves, flowers etc that are infected.
- But the disease severity is assessed by estimating the proportion of total photosynthetic area that is diseased.
- While this is generally less precise and less controllable than counting individual plants, it is usually a better predictor of crop loss.
- Because judging the proportion of diseased leaf by eye is unreliable, disease assessment keys, showing different disease severities as blackened areas, have been devised for various crops.

Crop Losses

- Once the amount of disease has been determined, the subsequently step is to assess, either experimentally or statistically, the effect of different levels of disease on the crop yield.
- Experimental assessment involves setting up test crops in which the level of disease is controlled.
- Monitoring crops with different levels of disease allows the comparison of epidemic progress and crop yield under different disease conditions.
- A relationship between disease parameters and yield can then be formulated, allowing prediction of crop loss for a certain level of disease at a particular point in the crop's growth.
- This method is dependent upon the assumption that the treatments used to keep disease at a certain level have no effect on crop yield themselves.

- Likewise, assessments using susceptible and resistant cultivars rely on both cultivars having a similar yield to start with.
- Again, this might not be true, and must be determined first in a disease-free environment.
- All aspects of experimental design need to be cautiously considered in order to gain an accurate assessment of the effect of disease on crop yield.
- For instance, harvesting methods under experimental conditions are often more efficient than under field conditions, giving the appearance of a higher crop yield.
- From crop loss assessment studies, models can be devised.
- Crop loss models are usually based on one of three types of disease assessment: disease at a critical point in development, disease at multiple points in development, and disease throughout crop development.
- Disease assessment methods will only be precise if performed on a representative sample of the crop.
- Samples of crop units (plants, leaves, fruit etc) can be taken randomly from a crop, or standard quadrats can be placed in the crop and all plants within the quadrat assessed.
- In a test plot, a part of the plot is generally assessed for disease.
- Taking samples only from the edge of a plantation will not necessarily be representative of the bulk of the crop.
- To determine how many samples need to be taken, it is probable to sample a number of times with progressively more samples, and find the point at which the standard error is low.
- A disease that is uniformly spread all through a crop will need fewer samples for a precise assessment than a disease that has a patchy distribution throughout the crop.

5

Diagnosis of Plant Pathogens

Detection/Diagnosis

Detection

- An art or skill of identifying some abnormalities, i.e. which is below or higher than the normal permissible range/threshold.
- Detection means the existence of whatever do you want detecting but diagnosis is that the analysis in depth of existing thing.
- Objectively look for, listen to the credible complains, symptoms that is sometimes not visible at all.

Diagnosis

- To ascertain the specific medical condition that is causing the problem.
- It a complex problem, where based on known theory, physiology, pathological findings, his experience one has to come to conclusion and answer to the question that "Why such symptoms exist?"
- It is a-priory step for prescribing medications for the plants.

How does a plant pathologist go about diagnosing plant problems?

1. Diagnostician must have very good observation skills.
2. Needs to be a good detective.
3. Keep an open mind until all of the facts related to the problem can be collected.
4. Possibility of multiple causal factors must also be considered.
5. Control measures depend on proper identification of diseases and of the causal agents.
6. Without proper identification of the disease and the disease causing agent, disease control measures can be a waste of time and money and can lead to further plant losses.
7. Proper disease diagnosis is therefore vital.
8. Hence, diagnosis is one of the most important aspects of a plant pathologist's training.

Techniques used for detection of pathogens

1. **Conventional**: On the basis of recognising symptoms, growth in culture, microscopy.
2. **Serological**: Interaction between antigen and antibody.
3. **Molecular techniques**: Nucleic acid based methods.

Conventional

1. Recognising symptoms
2. Growth in culture
3. Microscopy
4. Staining

I. Recognising symptoms

1. Symptoms of disease in infected plants can help us diagnose which pathogen is responsible.
2. Different pathogens can cause similar symptoms.
3. Ooze production is the characteristic symptoms of plant pathogenic bacteria.
4. It's usually necessary to do further diagnostic tests to accurately identify the pathogen causing the disease.

II. Growth in culture

1. Testing often starts by taking a sample from plant tissue showing symptoms of infection.
2. Cells from the sample are spread over a Petri dish that contains a growth medium such as PDA.
3. Growth medium contains nutrients.
4. Petri dish is kept in warm conditions in an incubator.
5. Bacteria and fungi from the sample grow to form cultures.
6. Appearance of a culture, such as its colour and furriness, can help a pathologist to identify the pathogen.
7. Different substances can be added to the growth medium to encourage or inhibit the growth of particular microorganisms.
8. This can help to identify the pathogens present in the sample.
9. Ex. *Xanthonas axonopodis pv. citri* gives yellowish color growth when streaked on growth media.
10. Ex.In fungi; Rhizopus species gives the cottony growth on PDA while, *Geotrichum candidum* gives the condensed growth.

III. Microscopy

1. Microsocpy detect the pathogen on the basis of morphology.
2. Bacteria, fungi and protists (but not viruses) can usually be observed using a light microscope.
3. Different pathogens have different shapes.
4. Some pathogens grow in pairs or chains. These characteristics can help us to identify them.
5. Different pathogens can be stained different colours using dyes.
6. One commonly used dye is Gram stain. This is another piece of evidence that can help us to identify them.
7. In case of fungi; reproductive structures, spore can be seen under microscope.
8. From which you can identify fungus upto genera level.
9. Nematodes can be detected after preparing slide from suspension, and visualized it under microscope.

IV. Staining

1. Bacterial cells, fungi and nematode can be colored with a stain.
2. Provide contrast with the background of to make cellular organalles visible.
3. Simple staining; methylene blue, basic fushchin and crystal violet
4. Steps for staining bacterial smear:
   - Pass the slide, smear the slide up through a flame three times.
   - Flood the smear with a stain and let stand one minute.
   - Rinse the slide gently with water, making sure the stream of water does not strike the smear directly.

Gram Staining

- Developed in 1884 by Danish physician Hans Christian Gram.
- Cornerstone of bacterial identification and the taxonomic division.
- Scientific technique of staining that is used to differentiate the species of bacteria into 2 main groups, namely the gram-positive and the gram-negative bacteria.
- The differences between these groups of bacteria are considered to be important in determining appropriate treatments for infections.

Serological Techniques

A method using the specificity of the antigen antibody reaction for the detection and identification of antigenic substances and the organism that carry them

1. Based on the specificity of antigen-antibody reactions.
2. Used for the diagnosis of plant viruses for many years .
3. Of late, these techniques have also been applied to diagnose fungi, bacteria and phytoplasmas.
4. Techniques originally based on the use of polyclonal antibodies (a mixture antibody), which often reduced the reliability of the test due to non specific cross reactions.
5. Development of monoclonal antibodies, which have narrow specificity against a single type of antigen, has greatly improved the accuracy of such tests.
6. Polyclonal antibody often react with several species within a genus, whereas monoclonal antibodies discriminate between different species within the genus.

Test kit

1. A sample from an infected plant can be put on a test strip that is coated with monoclonal antibodies.
2. Antibodies recognise an antigen from a specific pathogen.
3. Test strip works like a home pregnancy test.
4. Test line on the strip will change colour when the antigen from the pathogen is present.

Uses

1. To determine relationships between viruses .
2. To identify a virus causing plant disease.
3. To detect virus in foundation stocks of plants.
4. To detect symptomless viral infection.
5. To measure virus quantitatively.
6. To locate the virus within a cell or tissue.
7. To detect plant viruses in insects and to purify a virus

Enzyme-Linked Immunosorbent Assay (ELISA)

1. Molecular method for identification of diseases based on antibodies and color change in the assay.
2. In this method, the target antigens from the viruses, bacteria and fungi are made to specifically bind with antibodies conjugated to an enzyme.
3. The detection can be visualized based on color changes resulting from the interaction between the substrate and the immobilized enzyme.

4. Performance of ELISA can be improved greatly with the application of specific monoclonal and recombinant antibodies which are commercially available.
5. Specific monoclonal antibodies have been used in ELISA to achieve lower limits of detection in the order of 105–106 cfu/mL.
6. For plant disease detection, tissue print-ELISA and lateral flow devices that enable detection have been fabricated for on-site detection.
7. However, the sensitivity for bacteria is relatively low (105–106 cfu/mL), making it useful only for the confirmation of plant diseases after visual symptoms appear but not for early detection before disease symptoms occur.

Serology of Plant Viruses

1. Antigen: A virus protein or any other foreign protein.
2. When antigen injected into a mammal (rabbit, mouse, horse) or bird (chicken, turkey), it induces animal to produce specific new protein called Antibodies.
3. Antibodies circulate in the blood fluid or serum of animal.
4. Antibodies react specifically with and bind to a small portion of the surface called Antigenic determinant of the antigen injected.
5. Each antigen has many different antigenic determinants at its surface and produces antiserum.
6. An antigen producing cell is fused with a mouse myeloma (cell), it produces a hybrid cell that can grow in culture indefinitely and continues to produce monoclonal antibodies for a long time.
7. Such antibody producing hybrid cell is called hybridomas.
8. Hybridomas can be grown in culture for months or years and produce large quantities of identical monoclonal antibodies.

Monoclonal Vs Polyclonal antibodies

Antiserum: Serum containing antibodies

- **Monoclonal**
  - Antiserum contain single antibody.
  - Each antibody react only with single antigen of pathogen.
- **Polyclonal**
  - Antiserum contains a mixture of different antibodies.
  - Each antibody reacting with the antigen but at a different surface locality.

Immunofluorescent Staining

1. Identification of virus is further facilitated by coating of virus particles already on grid with antibodies (decoration) that make them appear quite distinctive under the electron microscope.
2. Part of plant leaves, whole cells, or cell secretions are first fixed that is killed with acetone or other organic compounds.
3. Fixed leaves tissues are then treated with antibodies to a virus that had been previously labelled with a compound, such as fluorescin isotiocyanate (FITC) which fluoresces under ultraviolet light.
4. If treated cells are infected with virus, the virus traps the antibodies and the attached fluorescent compound.
5. When such cells are viewed with a microscope supplied with ultraviolet light , cells or cell parts that contain virus appear fluorescent while the rest of the cells or cell areas appear dark.

Immunosorbent Electron Microscopy

1. Used for identification of virus present in low concentration or in mixture with other viruses.
2. EM Grids first coated with antibody to the target virus.
3. Then virus sample is placed on the antibodycoated grid.
4. Antibodies trap the virus from the sample and concentrate it on the grid where it can be easily found with the electron microscope.

Molecular Technique

Any properties of an individual's showing heritable variations is referred to as a character.

These characters which can be easily identified are referred to as marker.

Types of markers

Sl.No	Name of marker	Remarks
1	Morphological Markers	• Markers related to size, color, and surface of the various parts. • Such characters are used for varietal identifications
2	Biochemical Markers	• Markers related to variations in protein and amino acids banding patterns. • Gel electrophoresis studies are used for identification of biochemical markers
3	Molecular markers	• A unique sequence of nucleotides found on a strand of DNA. • Markers related to variations in DNA fragments generated by restriction endonuclease enzymes. • DNA markers are also used at molecular markers • Also called genetic markers or DNA Markers

Polymerase Chain Reaction (PCR)

1. Also called molecular photocopying
2. PCR technique was invented by Kary B.Mullis in 1985.
3. It is a molecular biology technique for enzymatically replicating DNA without using living organisms like *E.coli*, Yeast.
4. PCR are used to amplify or make many copies of small fragment of DNA.
5. PCR process is directed by a machine called thermocycler.
6. The machine is programmed to alter the temperature of the reaction every few minutes to allow DNA denaturing and synthesis
7. Taq polymerase enzyme, which is tolerant to high temperature is used in the PCR process.
8. Taq polymerase was discovered in hot springs from the bacteria *Thermus aquaticus.*
9. DNA polymerase is stable at high temperatures needed to perform the amplification.

Procedure of PCR

Each cycle of PCR consists of three important steps

Denaturing

1. Achieved at high temperature (94-96^{0}C).
2. Makes both templates DNA primers single stranded.
3. Takes usually 1-2 minutes.

Annealing

1. Achieved by lowering the temperature (45-60^{0}C).
2. Leads to attachment of primers (single stranded) to single DNA strands.

Elongation

1. Allows DNA polymerase to copy to DNA strand.
2. Final elongation step is used after the last cycle.

Application

1. Repeated heating and cooling cycles make several copies of the target DNA.
2. Twenty PCR cycles can be completed and millions of copies of the target DNA can be made in about one hour.
3. Based on the fidelity of DNA hybridization and replication, PCR was initially used for highly specific detection of diseases caused by bacteria

and viruses. Now it has been widely used for the detection of plant pathogens as well.

4. Reverse-transcription PCR (RT-PCR) has also been used for plant pathogen identification due to its high sensitivity.
5. Multiplex PCR was proposed to enable simultaneous detection of different DNA or RNA by running a single reaction .
6. Real-time PCR platforms have also been used for on-site, rapid diagnosis of plant diseases based on the bacterial, fungal and viral nucleic acids.

Advantages of PCR

1. Fast and inexpensive techniques.
2. It permits amplification of the targeted segment of DNA.
3. Used to clone gene of interest.
4. Performed with small amount of DNA.
5. *In vitro* multiplication of DNA.
6. Widely used by researchers.

PCR based method for detection of plant pathogen

1. Conventional PCR
2. Nested PCR
3. Multiplex PCR
4. Reverse Transcription-PCR
5. Real time PCR

1. *Conventional PCR*

- Basic method for the identification of fungal pathogens at different taxonomic levels.
- Taxonomic levels like genus, species or strain, depending on the specificity of the primers.
- Examples: Identified *Sclerotium rolfsii* and *Colletotrichum capsici* (ITS region) by conventional PCR method

2. *Nested PCR*

- Nested PCR is a variation of the PCR.
- Two pairs (in stead of one pair) of PCR primers are used to amplify a fragment.
- First pair of PCR primer amplify a fragment similar to a standard PCR.
- Second pair of primers called nested primers bind in side the first PCR product fragment.

- It allow amplification of a second PCR product which is shorter than the first one.
- This method has been widely used for detection and characterization of numerous fungi.

3. ***Multiplex PCR***
   - Used when more than two primer involve in the same reaction.
   - It allow simultaneous and sensitive detection of different DNA targets.
   - Highly useful for the identification of plant pathogen since plants are usually infected by more than one pathogen.
   - Used for the simultaneous detection and differentiation of
     - *Podosphaera xanthii* and *Golovinomyces cichoracearum* in sunflowe.
     - For detecting *Phytophthora lateralis* in cedar trees and water samples.

4. ***Reverse Transcription-PCR***
   - Used to amplify RNA sequence in DNA duplex.
   - In this technique, cDNA copy of the RNA is produced using the enzyme reverse transcriptase.
   - This cDNA is then used for amplification.
   - m-RNA is degraded rapidly in dead cells, the detection of mRNA by RT-PCR is considered an accurate indicator of cell viability.
   - RT-PCR has been used to detect viable population of *Mycosphaerella graminicola* in wheat.
   - This approaches highly valuable in phytopathology for the analysis of fungal gene expression during disease development.

5. ***Real time PCR***
   - Most sensitive method.
   - Discriminate between closely related mRNAs.
   - Developed to quantify differences in m RNA production, particularly in such cases where only small amounts of cells/tissue are available.
   - Various methods of mRNA quantitation include northern hybridization, ribonuclease protection assays, situ hybridization, and RT-PCR.
   - Conventional PCR does not yield truly and accurate quantitative result because ethidium bromide does not give a bright fluorescence, and when a band is detectable the logarithmic phase of amplification is already over.
   - Real time PCR is capable to perform multiplex detection of two or more pathogens in the same reaction.

6

Southern, Northern and Western Blotting Technique

What is Blotting

Techniques of transforming

1. DNA
2. RNA
3. Proteins

On to a carrier so they can be separated and often follows the use of a gel electrophoresis

Types of Blotting Technique

Three types

1. Southern Blot : Used to detect the DNA.
2. Northern Blot : Used to detect RNA.
3. Western Blot : Used to detect Proteins.

Southern blotting

- Used to detect presence of a specific DNA sequence.
- Technique developed by Edwin Southern in 1975.
- Also known as Southern Blot, Southern Transfer, Southern Hybridization and DNA gel blot analysis.
- Involves agarose gel electrophoresis.
- Involves capillary transfer.
- Uses DNA probe.

Application of Southern blotting

- Gene discovery
- Diagnostics
- Mapping
- Evolution
- Forensics

Northern blotting

- Used to determine the presence of a specific mRNA from a mixture of RNA.
- Technique was developed by Alwine and coworkers in 1977 and further elaborated by Thomas in 1980.
- Also known as Northern Plot, Northern Transfer and RNA Gel Blot Analysis.
- Involves denaturing formaldehyde agarose gel .
- Involves capillary transfer.
- Uses c DNA probes.

Application of Northern blotting

- Direct study of gene expression at the level of m RNA.
- Detection of m RNA transcript size.
- Study of RNA degradation.
- Study of RNA splicing.
- Study of RNA half life.

Western Blotting

- Used to detect the presence of a specific protein in a sample by using antibody specific to that protein.
- Technique developed by Ipwbin and coworkers in 1979.
- In this technique an antibody is used to detect a specific protein.
- Involves SDS page.
- Involves electric transfer.
- Use primary and secondary antibodies.
- Used in disease diagnosis.

Application of Western blotting

- To identify specific proteins.
- To identify their masses.
- Conformity of HIV test.

7

Polymerase Chain Reaction (PCR)

Polymerase Chain Reaction (PCR)

1. Also called molecular photocopying.
2. PCR technique was invented by Kary B.Mullis in 1985.
3. It is a molecular biology technique for enzymatically replicating DNA without using living organisms like E.coli, Yeast.
4. PCR are used to amplify or make many copies of small fragment of DNA.
5. PCR process is directed by a machine called thermocycler.
6. The machine is programmed to alter the temperature of the reaction every few minutes to allow DNA denaturing and synthesis.
7. Taq polymerase enzyme, which is tolerant to high temperature is used in the PCR process.
8. Taq polymerase was discovered in hot springs from the bacteria *Thermus auquaticus*.
9. DNA polymerase is stable at high temperatures needed to perform the amplification.

Procedure of PCR

Each cycle of PCR consists of three important steps

Denaturing

1. Achieved at high temperature (94-96^{0}C).
2. Makes both templates DNA primers single stranded.
3. Takes usually 1-2 minutes.

Annealing

1. Achieved by lowering the temperature (45-60^{0}C).
2. Leads to attachment of primers (single stranded) to single DNA strands.

Elongation

1. Allows DNA polymerase to copy to DNA strand.
2. Final elongation step is used after the last cycle.

Application

1. Repeated heating and cooling cycles make several copies of the target DNA.
2. Twenty PCR cycles can be completed and millions of copies of the target DNA can be made in about one hour.
3. Based on the fidelity of DNA hybridization and replication, PCR was initially used for highly specific detection of diseases caused by bacteria and viruses. Now it has been widely used for the detection of plant pathogens as well.
4. Reverse-transcription PCR (RT-PCR) has also been used for plant pathogen identification due to its high sensitivity.
5. Multiplex PCR was proposed to enable simultaneous detection of different DNA or RNA by running a single reaction.
6. Real-time PCR platforms have also been used for on-site, rapid diagnosis of plant diseases based on the bacterial, fungal and viral nucleic acids.

Advantages of PCR

1. Fast and inexpensive techniques.
2. It permits amplification of the targeted segment of DNA.
3. Used to clone gene of interest.
4. Performed with small amount of DNA.
5. *In vitro* multiplication of DNA.
6. Widely used by researchers.

PCR based method for detection of plant pathogen

1. Conventional PCR
2. Nested PCR
3. Multiplex PCR
4. Reverse Transcription-PCR
5. Real time PCR

1. *Conventional PCR*

- Basic method for the identification of fungal pathogens at different taxonomic levels.
- Taxonomic levels like genus, species or strain, depending on the specificity of the primers.
- Examples : Identified *Sclerotium rolfsii* and *Colletotrichum capsici* (ITS region) by conventional PCR method.

2. *Nested PCR*

- Nested PCR is a variation of the PCR.
- Two pairs (in stead of one pair) of PCR primers are used to amplify a fragment.
- First pair of PCR primer amplify a fragment similar to a standard PCR.
- Second pair of primers called nested primers bind in side the first PCR product fragment.
- IT allow amplification of a second PCR product which is shorter than the first one.
- This method has been widely used for detection and characterization of numerous fungi.

3.*Multiplex PCR*

- Used when more than two primer involve in the same reaction.
- It allow simultaneous and sensitive detection of different DNA targets.
- Highly useful for the identification of plant pathogen since plants are usually infected by more than one pathogen.
- Used for the simultaneous detection and differentiation of
- *Podosphaera xanthii* and *Golovinomyces cichoracearum* in sunflower.
- For detecting *Phytophthora lateralis* in cedar trees and water samples.

4. *Reverse Transcription-PCR*

- Used to amplify RNA sequence in DNA duplex.
- In this technique, cDNA copy of the RNA is produced using the enzyme reverse transcriptase.
- This cDNA is then used for amplification.
- m-RNA is degraded rapidly in dead cells, the detection of mRNA by RT-PCR is considered an accurate indicator of cell viability.
- RT-PCR has been used to detect viable population of *Mycosphaerella graminicola* in wheat.
- This approaches highly valuable in phytopathology for the analysis of fungal gene expression during disease development.

5. *Real time PCR*

- Most sensitive method.
- Discriminate between closely related mRNAs.
- Developed to quantify differences in m RNA production, particularly in such cases where only small amounts of cells/tissue are available.

- Various methods of mRNA quantitation include northern hybridization, ribonuclease protection assays, situ hybridization, and RT-PCR.
- Conventional PCR does not yield truly and accurate quantitative result because ethidium bromide does not give a bright fluorescence, and when a band is detectable the logarithmic phase of amplification is already over.
- Real time PCR is capable to perform multiplex detection of two or more pathogens in the same reaction.

8

ELISA & ISEM

Types of Serological Tests

1. Precipitin test
2. Gel diffusion test
3. Agglutination reaction
4. Elisa (Direct and Indirect)

1. Precipitin test

- Earliest and quite common method.
- Antigena and antibody are mixed in a solution.

2. Gel diffusion test

- Antigen & Antibody diffuse toward one another through agar gel and meet in a zone in suitable concentrations.

3. Agglutination reaction

- When antigen is adsorbed on the surface of a large particle such as cell, plastid etc and these are precipitated by addition of antibodies.

In all these tests the reaction of antigen and antibody becomes visible either by precipitation of the two on the bottom of the test tube or by formation of a band at the interface where the two meet.

Enzyme-Linked Immunosorbent Assay (ELISA) -

1. Molecular method for identification of diseases based on antibodies and color change in the assay.
2. In this method, the target antigens from the viruses, bacteria and fungi are made to specifically bind with antibodies conjugated to an enzyme.
3. The detection can be visualized based on color changes resulting from the interaction between the substrate and the immobilized enzyme.
4. Performance of ELISA can be improved greatly with the application of specific monoclonal and recombinant antibodies which are commercially available.

5. Specific monoclonal antibodies have been used in ELISA to achieve lower limits of detection in the order of 10^5–10^6 cfu/ml.
6. For plant disease detection, tissue print-ELISA and lateral flow devices that enable detection have been fabricated for on-site detection.
7. However, the sensitivity for bacteria is relatively low (10^5–10^6 cfu/ml, making it useful only for the confirmation of plant diseases after visual symptoms appear but not for early detection before disease symptoms occur.

Types of ELISA

Two types

1. Direct ELISA
2. Indirect ELISA

1. Direct ELISA (Double antibody sandwich ELISA)

- The wells (capacity 0.4 ml) of a polystyrene microtiter plate are first filled with and then emptied of, sequentially,

a. Antibodies to the virus.

b. Virus preparation or sap from an infected plant.

c. Antibodies to the virus to which molecules of a particular enzyme have been attached.

d. A substrate for the enzyme, that is, a substance that the enzyme can break down and cause change in its color.

  - The substrate is not emptied but is kept in the well.
  - Within 30 or 60 minutes, the wells are read either visually or preferably with a colorimeter that measures the amount of color in each well.
  - Presence of color in the well indicates that there was virus in the sample (step b above).
  - Degree of visible coloration or the size of the reading given by the colorimeter is proportional to and therefore a measure of the amount of virus present in the sample.

2. Indirect ELISA

- Sequence of steps b and a is reversed.
- The wells (capacity 0.4 ml) of a polystyrene microtiter plate are first filled with and then emptied of, sequentially,

  a. Virus preparation or sap from an infected plant.

  b. Antibodies to the virus.

c. Antibodies in the antibody-enzyme complex are not those against the virus, they are antibodies against the antibody proteins of the animal in which the virus antibody are produced. (They are anti- rabbit antibodies produced in still in another animal, such as goat)

Advantages of indirect ELISA

- The same goat anti rabbit antibody enzyme complex can be used in steps c for assays for any virus, as long as the first antibodies, that is those against the virus used in step b, were produced in rabbit.

Advantages of ELISA

1. Sensitive.
2. Large number of sample can be tested concurrently.
3. Only small amount of antisera is required.
4. Results are quantitative.
5. Procedure can be semiautomated.
6. Assay can be run regardless of virus morphology and virus concentration.

ELISA has become one of the most popular serodiagnostic techniques especially for multiple samples because of this advantages.

Immuno-sorbent Electron Microscopy (ISEM) –

- Immuno-sorbent electron microscopy.
- Developed by Derrick (1973).
- Technique which combines the specificity of serological test with the possibility to visualize the types of viral antigen in the electron microscope.
- This technique not only helps in the detection of virus particles but also facilitates virus identification.

Methodology of ISEM

- First make use of a trapping of antiserum-specific virus particles onto EM grids that have been coated by the specific antiserum.
- Samples that are serologically related to the antibodies on the coated grids get contaminated on the grids enabling the detection of virus is samples.
- To examine the virus in isolation with the electro microscope, the sample is mounted on a grid covered with a thin film of plastic for instance formvar.

- These grids are copper discs with 3 mm diameter, containing a number of apertures -150 to 400 mechses /inch.
- The virus particles stick to the formvar, contrast can be improved by shadow casting or negative staining.
- The antigen antibody binding is visualized after negative staining of the complex.

9

Laboratory Equipment and Their Use

Tools & Implement

Tool

1. Simple device.
2. Generally hand operated.
3. Handling requires more skill than handling of implement.
4. Used to perform a definite function.
5. Examples are carpenter's tools, Wool: smith's tools, goldsmith's tools.

Implement

1. Used to perform a specific task.
2. Operated by hand, bullocks or machines.
3. Handling requires lesser skill than handling of tools.
4. Does not record any observation.

Equipment & Instrument

Equipment

- Items needed for a particular purpose such as research equipments, kitchen equipments.
- Used for a particular purpose.
- Generally power operated.
- Handling requires specifc skill.

Instrument

- Device used to perform a particular task.
- Generally a mechanical or electronic device.
- Power operated.
- Used for recording scientific observations.
- Required for precision work.
- Handling requires proper training.

- Used to measure instruments are pressure, height, sound and for other scientific measurements.

Generally, the terms equipments and instruments are used as synonyms.

Instrumentation –

- Collective use of instruments in the research laboratory is called instrumentation.
- Pathological instrumentation refers to use of various instruments or collective use of instruments for pathological studies.

Common Instruments Used in Plant Pathology –

- There is a long list of instruments that are used in Plant Pathology.
- The list of commonly used instruments in these disciplines is presented below:

1	Autoclave	2	Hot air oven
3	Laminar flow	4	Spectrophotometer
5	Electrophoresis	6	Light and electron microscopy
7	Incubator	8	Microtome
9	Centrifuge	10	Tissue homogenizer
11	Vortex mixer	12	Orbital shaker
13	Freeze dryer	14	Nano drop
15	GC-MS	16	HPLC
17	Thermocycler	18	UV Trans illuminator
19	Gene Gun	20	ELISA
21	DNA Sequence		

1. Autoclave

1. An autoclave is a pressure chamber used to sterilize equipment.
2. Used for the sterilization of culture media, autoclavable containers, plastic tubes, and pipette tips.
3. May be either horizontal or vertical.
4. Steam pressure is hydraulically maintained which helps to shoot up the temperature quickly.
5. Generally the steam sterilization is done at 15 p.s.i for 15 minutes.
6. Moist heat sterilization is performed in Autoclave under pressure. It has more penetrating power than dry heat
7. Most media can sterilized by heating at 121°C or 15 p.s.i pressure for 15-20 minutes.

Pressure [p.s.i.][lb/in²]	Temperature [°C]
5	107
6	110
10	115
15	121
20	126
25	130
30	135

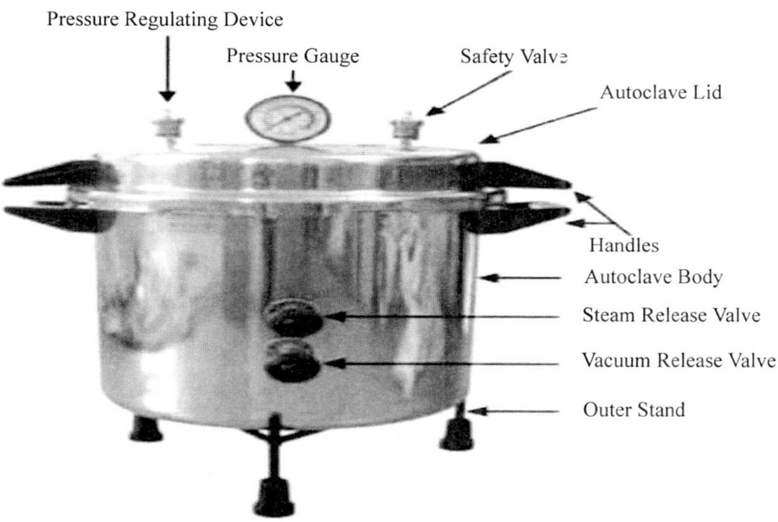

2. Hot Air Oven

- Vertical steel box with double or triple walled body of aluminium or stainless steel, positioning with wire mesh trays inside.
- **Major Parts:** Motorized blower, Thermostat control, Door, Digital temperature controller cum indicator etc.
- **Uses**; commonly used for dry heat sterilization of many objects such as glass wares like Petri plates, pipettes, flasks and other useable without graduations.
- Temperature required for sterilization

Temperature °C	Time
120	8 hr
140	3 hr
160	1 hr
180	20 min

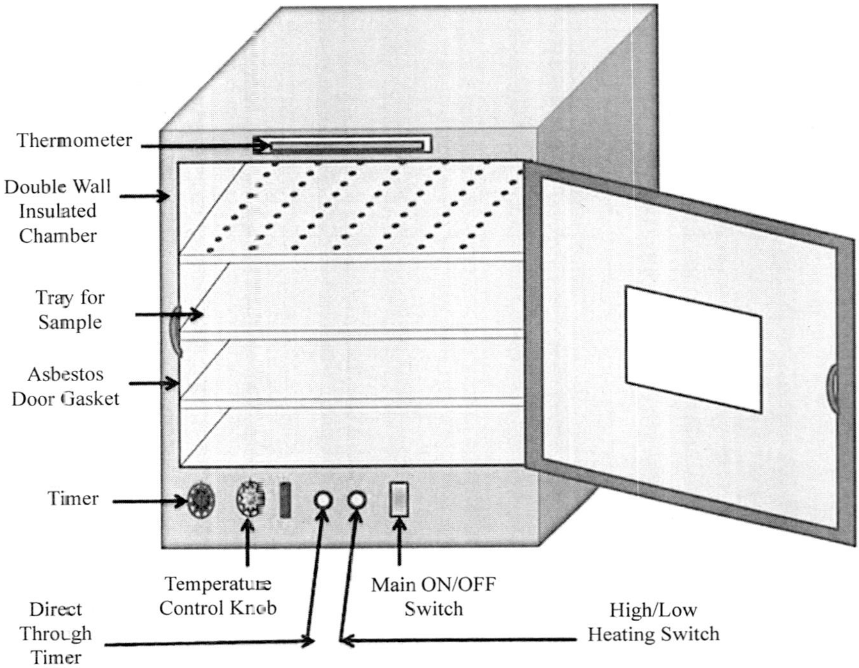

3. Laminar Air Flow

- A cabinet is fabricated out of thick board or is of stainless steel. Interior surface of working platform/ table is of stainless steel with the sunmicas clad at the top.
- Major parts: HEPA filter, Blower and motor
- Uses: provides an ascetic or microorganism free environment for performing various activities such as
- Pouring of sterilized media in sterilized plates.
- Isolation and transfer of pathogens during application of different methodologies require ascetic or sterilized environment.

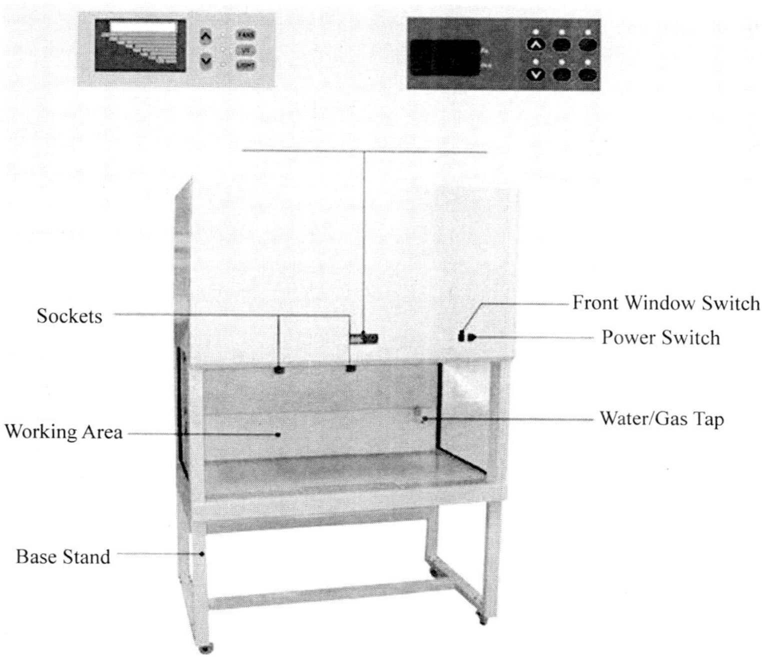

4. Spectrophotometer

- Used in laboratory for determining the concentration of substances in a solution allowing a qualitative or quantitative analysis of the sample.

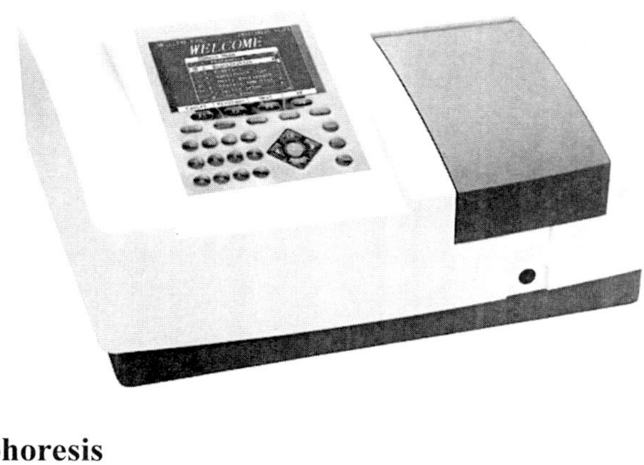

5. Electrophoresis

- Movement of charged particles (ions) in electric field resulting in their migration towards the oppositely charged electrode is known as electrophoresis.

- Used of separation of plasma proteins, lipoproteins and immunoglobulins.
- Common laboratory tools in the analysis of protein, DNA, RNA.
- Serum proteins is separated by gel electrophoresis to about 15 bands.
- Immuno electrophroresis is useful for analysis of complex mixture of antigen and antibodies.

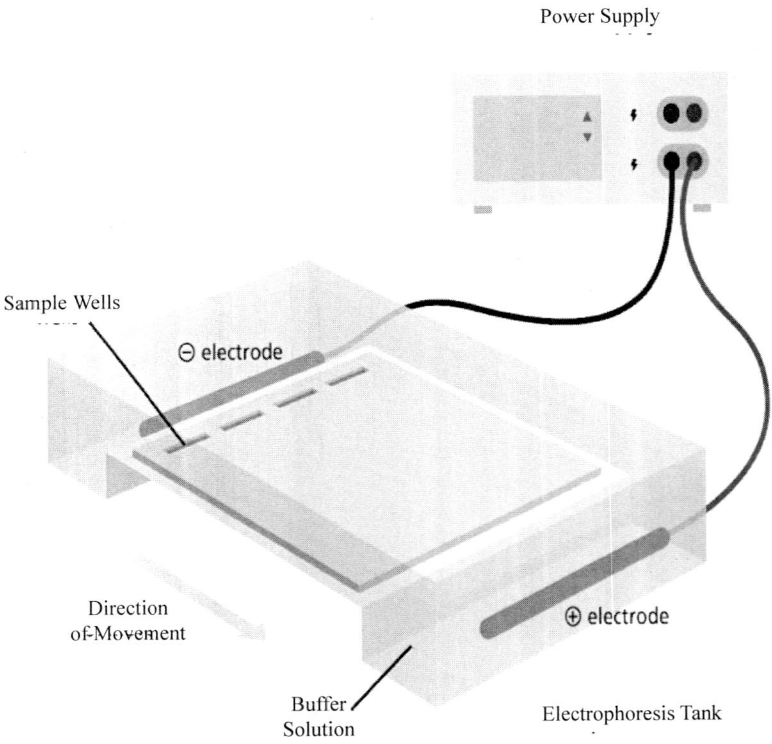

6. BOD Incubator

- BOD means Biological Oxygen Demand.
- Incubator is like a Vertical steel chamber shaped as an Almirah.
- Temperature inside may be maintained from 5 to 50°C.
- Incubators is provided with both heating and cooling systems.
- Used to maintain a range of temperature below and above the ambient temperature required for the growth and multiplication of various microorganisms.

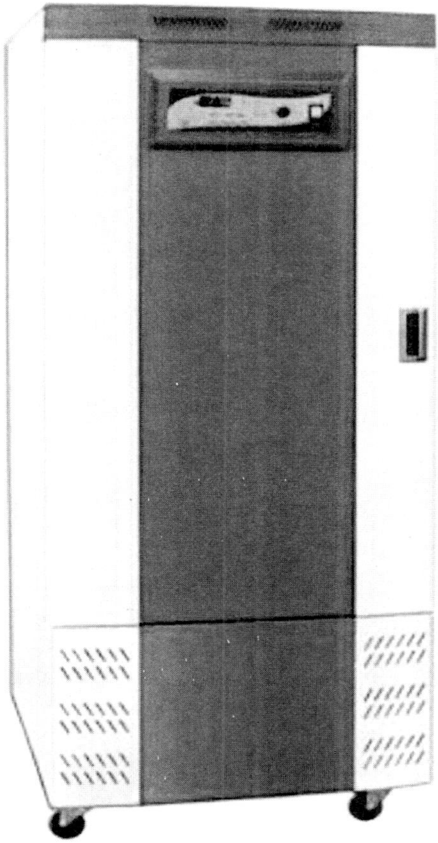

7. Microscope

- An instrument which is used for viewing small objects that cannot be seen with naked or unaided eye.
- Science of investigating small objects using microscope is called microscopy.
- Microscopic refers to minute or very small objects (things) which are not microscopy- visible with the unaided eye.
- Microscopic objects are examined with the help of microscope.

Types

- There is a variety of microscopes.
- Following five types of microscopes are very commonly used in plant pathology

1. Compound microscope
2. Stereo-microscope
3. Confocal microscope
4. Scanning Electron Microscope (SEM)
5. Transmission Electron Microscope (TEM)

Compound Microscope:

1. Most commonly used microscope
2. Light illuminated.
3. Image seen with this type of microscope is two dimensional.
4. Can be used to view individual cells, even living ones.
5. Has high magnification (4X — 100X).
6. Has a low resolution.
7. Visible light is used as source of radiation for image formation.
8. Air is used as medium.
9. Glass slides are used for mounting the specimen.
10. Focusing is done mechanically.
11. Specimen contrast is provided through light absorption. provides clear images of cytological preparations such as cell division (mitosis and meiosis).

Stereo Microscope

- Known as dissection microscope because as used for dissection purpose.
- Light illuminated.
- Image that appears is three dimensional.
- Cannot be used to view individual cells because of low magnification.
- Visible light is used for image formation.
- Air is used as medium.
- Tissues are viewed directly.
- Glass lenses are used.
- Focusing is done mechanically.
- Contrast is provided by light reflection.
- Used to view bigger objects.

Confocal Microscope

1. Microscope uses laser light.
2. Laser light scan across the specimen with the aid of sing mirrors.
3. Image is then placed on a digital computer screen for analyzing.
4. Uses laser light.
5. Air is used as medium.
6. Use of Glass slides with dyed samples.
7. Use of Glass lenses with dichromatic mirrors.
8. Focusing is digital computerized.
9. Magnification is digitally enhanced.

Scanning Electron Microscope (SEM)

- Electron illuminated.
- Image is seen in three dimensions.
- High magnification and high resolution.
- Specimen is coated in gold and the electrons bounce off to give an exterior view of the specimen.
- Pictures are in black and white.
- Vacuum is used as medium.
- Specimen is mounted on aluminium stubs and are coated in gold.
- One electrostatic lens and few electromagenetic lenses are used.
- Focusing is done electrically.
- Magnification adjustments are electrical.
- Specimen contrast is provided by electron scattering.

Transmission Electron Microscope (TEM)

- Electron illuminated.
- Three dimensional view.
- High magnification.
- Vacuum is used as medium.
- Lenses are like SEM.
- Specimen is mounted on thin films of colloid on copper grids
- Focusing is electrical.
- Magnification adjustments are electrical.
- Specimen contrast is provided by electron scattering.

8. Microtome

- Mechanical instrument used to cut biological specimens into very thin segment for microscopic examination.
- Most microtonics prepare sections of plant tissues for histology.
- The most common applications of microtomes are:

Traditional histological technique

- Tissues are hardened by replacing paraffin wax.
- Tissue is then cut in the microtome at thicknesses varying from **2 to 25** micrometers thick.
- From there the tissue can be mounted on a microscope slide, stained and examined using a light microscope.

Cryosection

- Water-rich tissues are hardened by freezing and cut frozen.
- Sections are stained and examined with a light microscope.
- Technique is much faster than traditional histology (5 minutes vs 16 hours).
- Used in operations to achieve a quick diagnosis.
- Cryosections can also be used in immuno histo chemistry.

Electron microscopy

- After embedding tissues in epoxy resin, a microtome equipped with a glass or diamond knife is used to cut very thin sections (typically 60 to 100 nanometers)
- Sections are stained and examined with a transmission electron microscope.
- This instrument is often called an ultramicrotome

Botanical microtomy

- Hard materials like wood, bone and leather require a sledge microtome.
- These microtomes have heavier blades and cannot cut as thin a regular microtomy.
- Microtome blades are extremely sharp, and should be handled with great care.
- Safety precautions should be taken in order to avoid any contact with the cutting edge of the blade.
- Microtomes are of two types, *viz.*, hand operated and power operated.

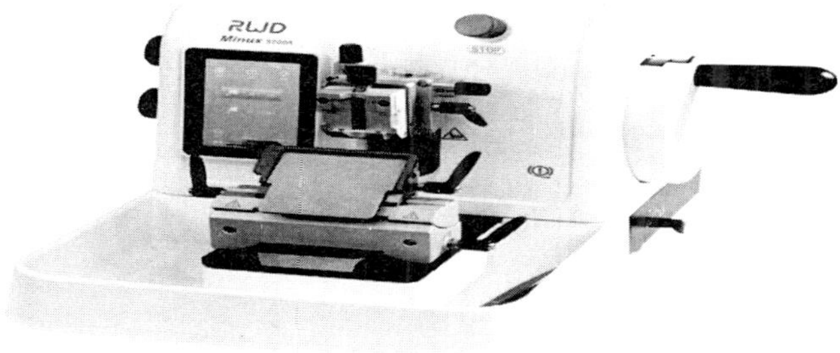

9. Centrifuge

- An equipment used to separate lighter and heavier particles from a substance.
- Centrifugal force is used for separation of particles.
- Sample is rotated around a fixed axis for separation into lighter and heavier particles.
- Rotation of the sample is driven by a motor.
- There are many different kinds of centrifuges such as simple centrifuge and ultra centrifuge.
- Simple centrifuges are used in biology and biochemistry for isolating and separating biocompounds on the basis of molecular weight.
- Ultracentrifuge has very high speed (1,000,000 G or 9,800 km/s2).
- There are two kinds of ultracentrifuges, viz., preparative and analytical.
- Both types have important applications in molecular biology.
- In biology, preparative ultracentrifuge is used for separation of cell organelles such as mitochondria, ribosomes, microsomes, DNA markers, etc.

10. Tissue Homogenizer

- Laboratory tools used to reduce the tissues into uniform sized fine molecules (0.2 to 2mm).
- Tissue homogenizers work better with moist tissues than ones dry.
- Homogenizers are used for following purposes.
    - Preparation of sample for PCR analysis.
    - Preparation of sample for DNA and RNA analysis.
    - Preparation of cell suspension.

Three types of homogenizers, *viz.*

1. Rotor Type

- Most common type of homogenizer.
- Very quick and efficient.
- Used to homogenize plant and animal tissues.
- Material is completely homogenized within 20 to 60 seconds.
- Foaming and aerosols are main problems with certain types of materials.

2. High Pressure Type

- Very high pressure is used to homogenize the material.
- Used to homogenize liquids such as milk and other such materials (fruit juices etc.)

3. Sonic Disruptors

- Known as sonicator.
- It reduces the material into microscopic particles through ultrasonic waves.
- Produce free radicals.
- Thus, it disrupts the sample into very fine particles.

11. Orbital Shaker

- Orbital shaker has a wide range of laboratory applications.
- Application include cell culture, tissue culture, hybridization (somatic), staining and destaining gels.
- Various types of orbital shakers are available such as fixed speed orbital shaker, variable speed orbital shaker (VSOS).
- Should have wide speed range.
- Should provide uniform orbital shaking motion.

- Should have heavy duty and durable.
- Should have adjustable bar for samples.

12. Gene Gun

1. Invented by John stanford in USA.
2. Device for injecting cells with genetic information.
3. Originally designed for plant transformation.
4. Payload is an elemental particle of a heavy metal coated with plasmid DNA.
5. Actual name of the gene gun is the Biolistic Particle Delivery Systems.
6. Often simply referred to as biolistics.
7. Gene gun was originally a nail gun for concrete surfaces modified to fire tungsten particles.
8. Later the design was greatly refined include the use of helium propellant and a multi-disk-collision delivery mechanism.
9. Other heavy metals such as gold and silver are also used.
10. It is used for DNA delivery in the host plant.

13. ELISA

- Biochemical technique used mainly in immunology to detect the presence of an antibody or an antigen in a sample.
- It utilizes two antibodies, one of which is specific to the antigen and the other of which is coupled to an enzyme.
- This second antibody gives the assay its "enzyme-linked" name, and will cause a chromogenic or fluorogenic substrate to produce a signal.
- ELISA can be performed to evaluate either the presence of antigen or the presence of antibody in a sample.
- It is a useful tool both for determining serum antibody concentrations (such as with the Human Immunodeficiency Virus, HIV test or West Nile Virus)
- Also, for detecting the presence of antigen.

14. Thermal Cycler

- Machine in which the Polymerase Chain 'Reaction (PCR) process is carried out.
- This machine heats and cools the reaction tubes within it to the precise temperature required for each step of the reaction.

- To prevent evaporation of the reaction mixture., a heated lid is placed on top of the reaction tubes or a layer of oil is put on the surface of the reaction mixture.

15. DNA Sequencer

- Machine for the automated determination of base sequences in a DNA sample.
- Very high accuracy in determining exact sequence of bases (adenine, guanine and cytosine) in a DNA fragment or a complete gene.
- Printout of DNA sequencing can be obtained from the DNA sequence.
- Colour printouts can also be obtained by using fluorescent nucleotides.
- Each DNA fragment (band) extracted through gel electrophoresis can be sequenced to show the exact order of bases.
- Entire genes are also sequenced, including DNA from chloroplasts, mitochondiria, introns, and genes that translate for large and small sub units of ribosomes.

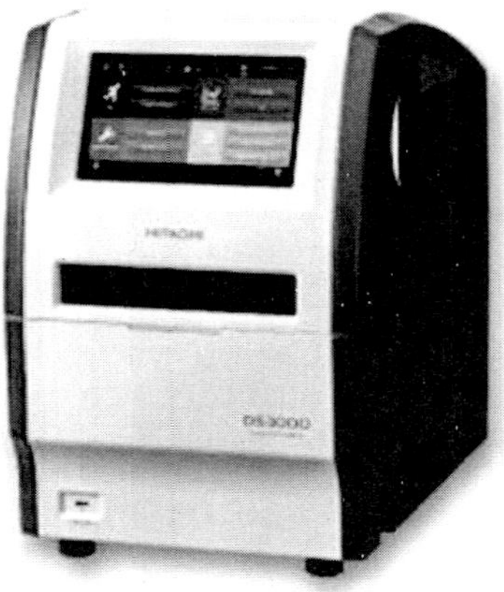

16. Freeze Dryer

- Freeze dryer removes water from a perishable material in order to preserve it, extending its shelf life and/or making it more convenient for transport.
- Freeze dryers work by freezing the material, then reducing the pressure and adding heat to allow the frozen water in the material to change directly to a vapor (sublimate).
- Freeze drying is also called lyophilization.
- A freeze dryer works in three phases:
  - Freezing
  - Primary drying (Sublimation)
  - Secondary drying (Adsorption)

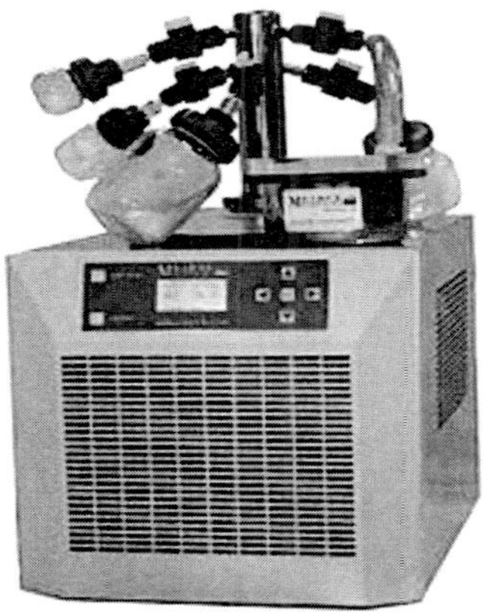

17.Nano Drop

- Common lab instrument that can measure the concentration of DNA, RNA, and protein.

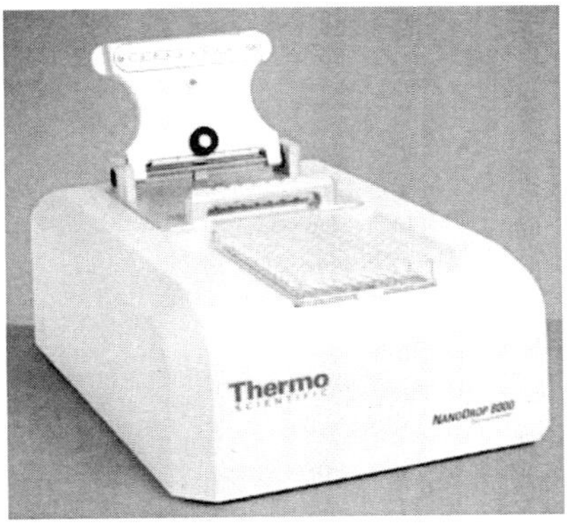

18. GC-MS

- GC-MS stand for Gas chromatography–Mass spectrometry (GC-MS).
- An analytical method that combines the features of gas-chromatography and mass spectrometry to identify different substances within a test sample.

Uses

- To study liquid, gaseous or solid samples.
- Identification of unknown samples.
- Detection of potential toxic chemicals in foods.
- Quantization of organic contaminants in water.

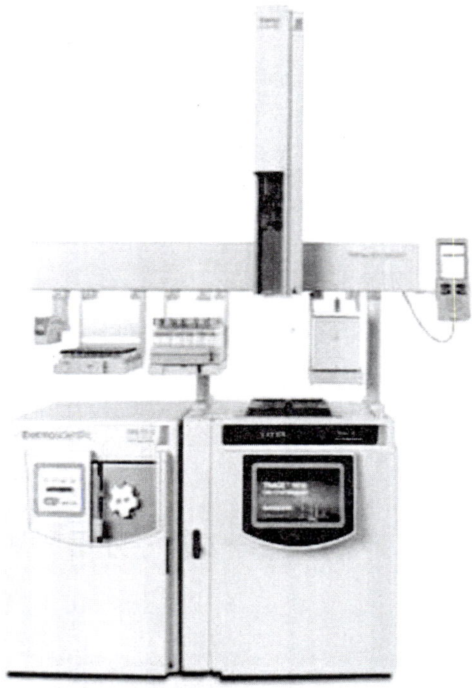

19. HPLC

- High-performance liquid chromatography (HPLC).
- Formerly referred to as high-pressure liquid chromatography.
- Chromatography can be described as a mass transfer process involving adsorption.
- Used to separate, identify, and quantify each component in a mixture.

20. UV Transilluminator

- Basic tool for visualizing flourescent stained gels.
- Used for viewing DNA in agarose gels stained with ethidium bromide.
- High and low intensity for proper gel visualization.
- UV blocking cover to protect user from hazardous UV radiation.
- Long life filter glass to inhibit solarization
- High sensitivity for visualization of even trace amounts of DNA.
- Stainless steel filter frame to reduce the change of gel damage or contamination.

21. Vortex Mixer

- Simple device used most commonly in cell culture and microbiology laboratories to suspend cells.
- Also used to mix many other substances.
- Consists of an electric motor with the drive shaft oriented vertically and attached to a cupped rubber piece mounted slightly off-center.
- As the motor runs the rubber piece oscillates rapidly in a circular motion.
- When a test tube or other appropriate container is pressed into the rubber cup (or touched to its edge) the motion is transmitted to the liquid inside and a vortex is created.
- Most vortex mixers have variable speed settings and can be set to run continuously, or to run only when downward pressure is applied to the rubber piece.

- Finger vortex" technique: An alternative to the electric vortex mixer.
- A technique in which a vortex is created manually by striking a test tube in a forward and downward motion with one's finger or thumb.
- This generally takes longer and often results in inadequate suspension, although it may be suitable in some cases when a vortex mixer is unavailable.
- Technique is better suited to accelerate the mixure of solutions which do not require the kinetic energy input needed to create suspensions.
- Used for vortexing cell suspensions, resuspending pellets, drug extraction, mixing tissue samples, enzymatic and RIA assays etc.

Unit 3: Physiological and Molecular Plant Pathology

1

Altered Metabolism of Plants under Biotic and Abiotic Stresses

1. Interferences of Plant Pathogens

Plant pathogens interfere in one or more ways with plant metabolism. These are;

1. Effect of pathogens on plant growth and reproduction.
2. Effect of pathogens on photosynthesis.
3. Effect of pathogens on translocation of water and nutrients in the host plant.
4. Effect of pathogens on translocation of organic nutrients through the phloem.
5. Effect of pathogens on host plant respiration.
6. Effect of pathogens on permeability of cell membranes.
7. Effect of pathogens on transcription and translation.

Effect of pathogens on plant growth

- Destroy part of the roots of a plant.
- Clog their xylem or phloem elements.
- Interference with the translocation of water. and of inorganic or organic nutrients.
- Destroy part of the photosynthetic area of plants.
- Reduced photosynthetic output.
- Smaller growth of these plants and smaller yields.
- Reduction in size and yields of plants and, sometimes, their death.

Examples

Some of the most common diseases in which pathogens cause obvious abnormal growth of their hosts' organs and tissues include:

- Clubroot of crucifers caused by the *Plasmodiophora brassicae*.

- Potato wart caused by the fungus *Spongospora subterranea.*
- Peach leaf curl caused by the fungus *Taphrina sp.*
- Corn smut caused by *Ustilago maydis.*
- Dwarf bunt of wheat caused by *Tilletia.*
- Crown gall of many hosts and hairy root of apple caused by *Agrobacterium tumefaciens and A. Rhizogenes.*
- Olive knot and oleander gall caused by *Pseudomonas savastanoi.*
- Leafy gall of several hosts caused by *Rhodococcus sp.*

Effect of pathogens on plant reproduction

Direct effects

- Many pathogens have a direct adverse effect on plant reproduction because they attack and kill the flowers, fruit, or seed directly, or interfere and inhibit their production.
- Pathogens interfere directly or indirectly with the propagation of their host plant.
- Example
    - Brown rot of stone fruits caused by the fungus *Monilinia sp.*
    - Bcterial canker and gummosis of stone fruit trees caused by *Pseudomonas syringae.*
    - Fireblight disease of pears and apples caused by the bacterium *Erwinia amylovora.*
    - Ergot of grains caused by the fungus *Claviceps purpurea.*
    - Corn smut and the covered and loose smuts of the various cereals caused by *Tilletia and Ustilago* sp., respectively.
    - Some diseases caused by viruses, phytoplasmas, or phloem-limited bacteria, no flowers are produced or those produced are sterile, and therefore few or no fruit and seed are produced.

Indirect effects

- Pathogens that attack various organs and tissues of plants weaken and often kill these organs or tissues, thereby weakening the plants.
- As a result, such plants remain smaller in size, may produce fewer flowers, and may set fewer fruit and seeds.
- Latter may be of inferior vigor and vitality.
- Therefore, if planted, they may produce fewer and weaker new plants.

2. Effect of Pathogens on Photosynthesis

- Pathogens do interfere with photosynthesis is obvious from the chlorosis, necrotic lesions, reduced growth and amounts of fruits produced by many infected plants.
- In leaf spot, blight, cereal rusts and fungal leaf spots, bacterial leaf spots, viral mosaics and yellowing and stunting diseases, or in defoliations, photosynthesis is reduced because the photosynthetic surface of the plant is lessened.
- The overall chlorophyll content of leaves in many fungal and bacterial diseases is reduced, but the photosynthetic activity of the remaining chlorophyll seems to remain unaffected.
- In some fungal and bacterial diseases, photosynthesis is reduced because the toxins, such as tentoxin and tabtoxin, produced by these pathogens inhibit some of the enzymes that are involved directly or indirectly in photosynthesis.
- In plants infected by many vascular pathogens, stomata remain partially closed, chlorophyll is reduced, and photosynthesis stops even before the plant eventually wilts.
- Most virus, mollicute, and nematode diseases also induce varying degrees of chlorosis and stunting, the photosynthesis of infected plants is reduced greatly.

3. Effect of pathogens on translocation of water and nutrients in the host plant

Many plant pathogens interfere in one or more ways with the translocation of water and inorganic nutrients through plants. These are :

3.1. Effect on absorption of water by roots.

3.2. Effect on translocation of water through the xylem.

3.3. Effect on transpiration

Effect on absorption of water by roots

- Some plant pathogens affect the integrity or function of the roots, causing them to absorb less water.
- They inhibit root hair production, alter the permeability of root cells etc that interferes with the normal absorption of water by roots.
- Examples of reduction of water absorption by plants
  - Destruction of roots of young seedlings by the damping-off oomycete *Pythium sp.*

- Roots and stems of pepper plants killed by *Phytophthora sp.*
- Wheat roots at different stages of destruction by the take-all fungus *Gaeumannomyces tritici.*
- Infection of crown and roots of corn plant with the fungus *Fusarium.*
- Numerous galls caused by the bacterium *Agrobacterium tumefaciens* on roots of a cherry tree.
- Root knot galls caused by the nematode *Meloidogyne sp.* on roots of a cantaloupe plant.

Effect on translocation of water through the Xylem

- Some pathogens, by growing in the xylem vessels or by other means interfere with the translocation of water through the stem.
- They interfere by clogging due to physical presence in the xylem as mycelium, spores, or bacterial cells, development of tyloses, induction of gall formation in the stem and roots etc.
- Examples of reduction of upward translocation of water and mineral nutrients by
    - Stem of a cantaloupe plant infected with the fungus *Phomopsis* sp.
    - Canker on an almond tree cause by the fungus *Ceratocystis fagacearum.*
    - Vascular wilt of tomato caused by the fungus *Fusarium.*
    - Discolored vascular tissues of a tomato stem infected with the same fungus.
    - Wilted tomato plants infected with the vascular bacterium *Ralstonia solanacearum.*

Effect on transpiration

- Some pathogens interfere with the water economy of the plant by causing excessive transpiration through their effects on leaves and stomata.
- This is the result of destruction of at least part of the protection afforded the leaf by the cuticle, an increase in the permeability of leaf cells, and the dysfunction of stomata etc.
- Ways by which pathogens cause increased transpiration in infected plants.
    - Wheat leaf rust pathogen *Puccinia recondita* produces innumerable lesions (uredia) on wheat leaves and causes millions of breaks in the leaf epidermis through which transpiration goes on uncontrollably.
    - Uredospores breaking the epidermis and emerging from the surface of an infected leaf.

- Grape berries infected with the powdery mildew fungus *Uncinula necator*, the mycelium of which penetrates and forms haustoria in almost every epidermal cell.
- Apple scab fungus *Venturia inaequalis* grows between the cuticle and the epidermis, causing the cuticle to break in numerous places, allowing transpiration to occur.
- Tomato leaves with numerous lesions caused by the fungus *Septoria sp.* and through which excessive transpiration occurs.

4. Interference with translocation of organic nutrients through the phloem

- Plant pathogens may interfere with the movement of organic nutrients from the leaf cells to the phloem, with their movement from the phloem into the cells that will utilize them.
- Obligate fungal parasites, such as rust and mildew fungi, cause an accumulation of photosynthetic products, as well as inorganic nutrients, in the areas invaded by the pathogen.
- In these diseases, the infected areas are characterized by reduced photosynthesis and increased respiration.
- In diseases caused by phytoplasmas, as well as in diseases caused by phloem-limited fastidious bacteria, bacteria exist and reproduce in the phloem sieve tubes, thereby interfering with the downward translocation of nutrients.
- In stem diseases of woody plants in which cankers develop, the pathogen attacks and remains confined to the bark for a considerable time.
- During that time the pathogen attacks and may destroy the phloem elements in that area, thereby interfering with the downward translocation of nutrients.

Examples of diseases in which the pathogen interferes with the downward translocation of organic nutrients

- Blister canker on a pine tree in which the bark and phloem have been killed by the fungus *Cronartium ribicola.*
- Phytoplasmas filling a phloem sieve element block the downward translocation of photosynthates.
- Potato tuber showing vein necrosis caused by the *potato leaf roll virus.*
- Young canker caused by the fungus *Nectria* in which the bark of the branch invaded and killed by the fungus.

- Two advanced *Nectria* cankers in which both the phloem and a great deal of the xylem have been killed by the fungus.

5. Effect of pathogens on host plant respiration –

- When plants are infected by pathogens, the rate of respiration generally increases.
- This means that affected tissues use up their reserve carbohydrates faster than healthy tissues would.
- The increased rate of respiration appears shortly after infection, certainly by the time of appearance of visible symptoms and continues to rise during the multiplication and sporulation of the pathogen. After that, respiration declines to normal levels or to levels even lower than those of healthy plants.
- Respiration increases more rapidly in infections of resistant varieties, in which large amounts of energy are needed and used for rapid production or mobilization of the defense mechanisms of the cells. In resistant varieties, however, respiration also declines quickly after it reaches its maximum.
- In susceptible varieties, in which no defense mechanisms can be mobilized quickly against a particular pathogen, respiration increases slowly after inoculation, but continues to rise and remains at a high level for much longer periods.
- The activity or concentration of several enzymes of the respiratory pathways seems to be increased.
- Increased respiration in diseased plants is also accompanied by an increased activation of the pentose pathway, which is the main source of phenolic compounds which play important roles in the defense mechanisms of the plant against infection.

5. Effect of pathogens on permeability of cell membranes –

- Cell membranes function as permeability barriers that allow passage into a cell only of substances the cell needs and inhibit passage out of the cell of substances needed by the cell.
- Changes in cell membrane permeability are often the first detectable responses of cells to infection by pathogens, to most toxins, to certain pathogen enzymes, and to certain toxic chemicals, such as air pollutants.
- Most commonly observed effect of changes in cell membrane permeability is the loss of electrolytes, i.e., of small water-soluble ions and molecules from the cell.

- Electrolyte leakage occurs much sooner and at a greater rate when the host–pathogen interaction is incompatible, and the host remains more resistant than when the host is susceptible and develops extensive symptoms.
- If pathogens affect cell membrane permeability directly, they bring this about by stimulating certain membrane- bound enzymes, such as ATPase, which are involved in the pumping of H− in and K+ out through the cell membrane, by interfering with processes required for the maintenance and repair of the fluid film making up the membrane, or by degrading the lipid or protein components of the membrane by pathogen produced enzymes.

6. Effect of pathogens on transcription and translation

Effect on Transcription

Effect on Translation

Effect on Transcription

- Several pathogens, particularly viruses and fungal obligate parasites, such as rusts and powdery mildews, affect the transcription process in infected cells.
- In few cases, pathogens affect transcription by changing the composition, structure, or function of the chromatin associated with the cell DNA.
- In some diseases, especially those caused by viruses, the pathogen, through its own enzyme or by modifying the host enzyme (RNA polymerase) that makes RNA, utilizes the host cell nucleotides and machinery to make its own (rather than host) RNA.
- In several diseases, the activity of ribonucleases (enzymes that break down RNA) is increased, perhaps by formation in infected plants of new kinds of ribonucleases not known to be produced in healthy plants.
- Lastly, in several diseases, infected plants, particularly resistant ones, seem to contain higher levels of RNA than healthy plants, especially in the early stages of infection.

Effect of pathogens on translation

- Infected plant tissues often have increased activity in several enzymes, particularly those associated with the generation of energy (respiration) or with the production or oxidation of various phenolic compounds, some of which may be involved in (defense) reactions to infection.

- Increases in protein synthesis in infected tissues have been observed primarily in hosts resistant to the pathogen and reach their highest levels in the early stages of infection, i.e., in the first few minutes and up to 2–20 hours after inoculation.
- If resistant tissues are treated before or during infection with inhibitors of protein synthesis, their resistance to the pathogen is reduced.
- These interpretation suggest that much of the increased protein synthesis in plants attacked by pathogens reflects the increased production of enzymes and other proteins involved in the defense reactions of plants.

2

Molecular Mechanisms of Pathogenesis

Pathogenesis

- Chain of events leading to the development of disease is called pathogenesis or disease cycle
- The primary events in a disease cycle are
  1. Inoculation
  2. Penetration
  3. Establishment of infection
  4. Colonization (Invasion)
  5. Growth and reproduction of pathogen
  6. Dissemination of pathogen
  7. Survival of pathogen

Inoculation

- Inoculation is the initial contact of a pathogen with a site of plant where infection is possible.
- Pathogen(s) that lands on or is otherwise brought into contact with the plant is called the inoculum.
- Inoculum is any part of the pathogen that can initiate infection.
- One unit of inoculum of any pathogen is called a propagule.
  - In fungi, the inoculum may be fragments of mycelium, spores, sclerotia (i.e., a compact mass of mycelium).
  - In bacteria, mollicutes, protozoa, viruses, and viroids, the inoculum is always whole individuals of bacteria mollicutes, protozoa, viruses, and viroids, respectively.
  - In nematodes, the inoculums may be adult nematodes, nematode juveniles, or eggs.
  - In parasitic higher plants, the inoculum may be plant fragments or seeds.

Types of Inoculum

- **Primary inoculums:** An inoculum that survives dormant in the winter or summer and causes the original infections in the spring or in the autumn is called a primary inoculum, and the infections it causes are called primary infections.
- **Secondary inoculums:** An inoculum produced from primary infections is called a secondary inoculum and it, in turn, causes secondary infections.

Sources of Inoculum

- **Fungi, bacteria, parasitic higher plants, and nematodes**
  - Either produce their inoculums on the surface of infected plants or their inoculums reaches the plant surface when the infected tissue breaks down.
- **Viruses, viroids, mollicutes, fastidious bacteria, and protozoa**
  - Produce their inoculum within the plants.
  - Such an inoculum almost never reaches the plant surface in nature.
  - Therefore, it can be transmitted from one plant to another almost entirely by some kind of vector, such as an insect.
- In some fungal and bacterial diseases of perennial plants, the inoculum is produced on the branches, trunks, or roots of the plants. other times it comes into the field with the seed, transplants, tubers, or other propagative organs or it may come from sources outside the field.
- In many plant diseases of annual crops, the inoculum survives in perennial weeds or alternate hosts, and every season it is carried from them to the annual and other plants.

Landing or Arrival of Inoculum

- Inoculum of most pathogens is carried to host plants passively by wind, water and insects.
- An airborne inoculum usually gets out of the air and onto the plant surface not just by gravity but by being washed out by rain.
- Some types of inoculum in the soil, e.g., zoospores and nematodes, may be attracted to the host plant by sugars and amino acids diffusing out of the plant roots.
- Vector transmitted pathogens are usually carried to their host plants with an extremely high efficiency.

Prepenetration Phenomena

1. Attachment of pathogen to host.
2. Spore germination and perception of the host surface.
3. Appressorium formation and maturation.
4. Recognition between host and pathogen.

Attachment of Pathogen to Host

Fungi, bacteria and parasitic higher plants

- They are first brought into contact with the external surface of plant organs.
- They must first become attached to the host surface before can penetrate and colonize the host.
- The propagules of these pathogens have on their surface or at their tips mucilaginous substances consisting of mixtures of water-insoluble polysaccharides, glycoproteins, lipids, and fibrillar materials, which, when moistened, become sticky and help the pathogen adhere to the plant.
- Powdery mildew fungi, do not require free water for infection, adhesion is accomplished by release from the spore of the enzyme cutinase, which makes the plant and spore areas of attachment more hydrophilic and cements the spore to the plant surface.
- Fungi causing vascular wilts, spores fail to adhere after hydration but become adhesive after they are allowed to respire and to synthesize new proteins.
- The act of attachment often seems essential for the subsequent transmission of signals for germ tube extension and production of infection structure.

Pathogens such as mollicutes, fastidious bacteria, protozoa, and most viruses

- They are placed directly into cells of plants by their vectors and they are probably immediately surrounded by cytoplasm, cytoplasmic membranes, and cell walls.

Spore germination and perception of the host surface

- Spores have mechanisms that prevent their germination until they sense such stimulations or when there are too many spores in their vicinity.
- Once the stimulation for germination has been received by the spore, the latter mobilizes its stored food reserves, such as lipids, polyoles,

and carbohydrates, and directs them toward the rapid synthesis of cell membrane and cell wall toward the germ tube formation and extension.

- Germ tube is a specialized structure often growing for a very short distance before it differentiates into an appressorium.
- The perception of signals from plant surfaces by pathogenic fungi seems to be the result of signaling pathways mediated by
  - Cyclic adenosine monophosphate (cAMP)
  - Nitogen-activated protein kinae (MAPK)

Cyclic adenosine monophosphate (cAMP)

- Transmission of the cAMP signal proceeds via the cAMP-dependent activity of protein kinase A (= PKA) and subsequent phosphorylation of target proteins.
- The major activity of PKA in developing germ tubes is the mobilization of carbohydrates and lipids to the appressorium site.
- Therefore, pivotal to the production of functional appressoria.

Nitogen-activated protein kinase (MAPK)

- Signaling pathways for infection-related development are also achieved through Nitogen-activated protein kinases (MAPKs) and their upstream regulatory kinases.
- All of these together comprise a functional unit that transmits input signals from the periphery of the cell to the cell nucleus to elicit the expression of appropriate genes.
- A MAP kinase, K1 or P1, regulates appressorium formation in response to a signal from the plant surface but it is also required for invasive growth or viability in its host plant.

Appressorium Formation and Maturation

- Once appressoria are formed, they adhere tightly to the leaf surface.
- Consequently, appressoria secrete extracellular enzymes, generate physical force or both to bring about penetration of the cuticle by the fungus.
- Appressoria must be attached to the host plant surface strongly enough to withstand the invasive physical force applied by the fungus and to resist the chemical action of the enzymes secreted by the fungus.
- The size of the turgor pressure inside an appressorium found to be 40 times greater than the pressure of a typical car tire.
- The turgor pressure of an appressorium is due to the enormous accumulation of glycerol in the appressorium, which, due to its high

osmotic pressure, draws water into the cell and generates hydrostatic pressure that pushes the thin hypha (appressorial penetration peg) outward through the host cuticle.

- Mobilization of spore-stored products to the developing appressorium and glycerol biosynthesis in it is regulated by the cAMP signaling pathway.
- Whereas the initial movement of lipid and glycogen reserves to the developing appressorium was also found to be regulated by the K1 MAP.
- Production of penetration hyphae appears to be regulated by a MAP kinase pathway.

Recognition between Host and Pathogen

- When a pathogen comes in contact with a host cell, an early event takes place that triggers a fairly rapid response in each organism that either allows or impedes further growth of the pathogen and development of disease.
- The nature of the "early event" is not known with certainty in any host–parasite combination.

Pathogen Components

Elicitors

- Released from attacking pathogens before or during entry into the host.
- They may have a narrow host range, e.g., the elicitins.
- Some elicitors may be components of the cell surface of the pathogen (e.g., b-glucans, chitin, or chitosan) that are released by the action of host enzymes (e.g., b-glucanase and/or chitinase) and have broad host ranges.
- Some may be synthesized and released by the pathogen after it enters the host in response to host signals.
- It include the harpin proteins of bacteria that induce development of the hypersensitive response, certain hydroxylipids, and certain peptides and carbohydrates that induce specific host defense responses such as the, production of phytoalexins.
- Elicitors are considered as determinants of pathogen avirulence, as by their presence they elicit the hypersensitive (resistance) response and initiation of transcription of the plant genes that encode the various components of the defense response.
- These defense measures by the host plant, in turn, result in the pathogen appearing as avirulent.

- When the initial recognition signal received by the pathogen favors growth and development, disease may be induced; if the signal suppresses pathogen growth and activity, disease may be aborted.
- However, if the initial recognition elicitor received by the host triggers a defense reaction, pathogen growth and activity may be slowed or stopped and disease may not develop; if the elicitor either suppresses or bypasses the defense reaction of the host, disease may develop.

Host components

- Acting as signals for recognition by and activation of pathogens .They may include
  - **Fatty acids:** parts of the plant cuticle that activate production by the pathogen of the cutinase enzyme, which breaks down cutin.
  - **Galacturonan:** molecules of host pectin, which stimulate the production of pectin lyase enzymes by the fungus or bacterium.
  - **Phenolic compounds:** such as strigol, which stimulate activation and germination of propagules of some pathogens.
  - **Isoflavones and other phenolics, amino acids, and sugars:** released from plant wounds that activate a series of genes in certain pathogens leading to infection.

Germination of Spores and Seeds

- Fungal spores first germinate by producing a typical mycelium that infects and grows into host plants or they produce a short germ tube that produces a specialized infectious structure, the haustorium .
- Some fungal spores germinate by producing other spores, e.g., sporangia produce zoospores and teliospores produce basidiospores.
- After spores germinate, the resulting germ tube must grow, or the motile secondary spore (zoospore) must move, toward a site on the plant surface at which successful penetration can take place .
- The number, length, and rate of growth of germ tubes, or the number and mobility of motile spores, may be affected by physical conditions, such as temperature and moisture, by the kind and amount of exudates the plant produces at its surface, and by the saprophytic microflora.
- Haustoria are also produced by many fungi.
- Seeds germinate by producing a radicle, which either penetrates the host plant directly or first produces a small plant that subsequently penetrates the host plant by means of specialized feeding organs called haustoria.
- Most conditions described earlier as affecting spore germination and the direction of growth of germ tubes also apply to seeds.

Penetration

- Pathogens penetrate plant surfaces by direct penetration of cell walls, through natural openings, or through wounds.
- Bacteria enter plants mostly through wounds, less frequently through natural openings, and never directly through unbroken cell walls .
- Viruses, viroids, mollicutes, fastidious bacteria, and protozoa enter through wounds made by vectors.
- Some viruses and viroids may also enter through wounds made by tools and other means.
- Parasitic higher plants enter their hosts by direct penetration.
- Nematodes enter plants by direct penetration and, sometimes, through natural openings.
- Penetration does not always lead to infection.

Direct Penetration through Intact Plant Surfaces

Fungi

- Probably the most common type of penetration by fungi, oomycetes, and nematodes.
- Only type of penetration by parasitic higher plants.
- None of the other pathogens can enter plants by direct penetration.
- Nonobligate parasitic or hemibiotrophic, penetrate through a *fine hypha* produced directly by the spore or mycelium .
- Obligately parasitic ones do so through a penetration peg produced by an appressorium.
- Fine hypha or appressorium is formed at the point of contact of the germ tube or mycelium with a plant surface.
- Fine hypha grows toward the plant surface and pierces the cuticle and the cell wall through mechanical force and enzymatic softening of the cell wall substances.
- Most fungi form an appressorium at the end of the germ tube, with a flat surface in contact with the surface of the host plant.
- Then, a penetration peg grows from the flat surface of the appressorium toward the host and pierces the cuticle and the cell wall.
- In most fungal diseases the fungus penetrates the plant cuticle and the cell wall, but in some, such as apple scab the fungus penetrates only the cuticle and stays between the cuticle and the cell wall.

Parasitic higher plants

- Form an appressorium and penetration peg at the point of contact of the radicle with the host plant, and penetration is similar to that in fungi.

Nematodes

- Direct penetration in nematodes is accomplished by repeated back-and-forth thrusts of their stylets.
- Such thrusts finally create a small opening in the cell wall.
- Nematode then inserts its stylet into the cell or the entire nematode enters the cell.

Penetration through Wounds

- All bacteria, most fungi, some viruses, and all viroids can enter plants through various types of wounds.
- Bacteria and fungi penetrating through wounds germinate or multiply in the wound sap or in a film of rain or dew water present on the wound.
- Subsequently, the pathogen invades adjacent plant cells or it secretes enzymes and toxins that kill and macerate the nearby cells.
- Some viruses and all mollicutes, fastidious vascular bacteria, and protozoa enter plants through wounds made by their vectors.
- All four types of pathogens are transmitted by certain types of insects.
- Some viruses are also transmitted by certain nematodes, mites and fungi.
- Some viruses and viroids are transmitted through wounds made by human hands and tools.

Penetration through Natural Openings

Many fungi and bacteria enter plants through stomata, hydathodes, nectarthodes, and lenticels.

Stomata

- Stomata are most numerous on the lower side of leaves.
- They open in the daytime but are more or less closed at night.
- Bacteria present in a film of water over a stoma and, if water soaking occurs, can swim through the stoma easily and into the substomatal cavity where they can multiply and start infection.
- Fungal spores generally germinate on the plant surface, and the germ tube may then grow through the stoma.
- Germ tube forms an appressorium that fits tightly over the stoma, and usually one fine hypha grows from it into the stoma.

Hydathodes

- Hydathodes are more or less permanently open pores at the margins and tips of leaves.
- Some bacteria use these pores as a means of entry into leaves, but few fungi seem to enter plants through hydathodes.

Nectarthodes

- Some bacteria also enter blossoms through the nectarthodes, nectaries, which are similar to hydathodes.

Lenticels

- Lenticels are openings on fruits, stems, and tubers that are filled with loosely connected cells that allow the passage of air.
- During the growing season, lenticels are open, few fungi and bacteria penetrate tissues through them, growing and advancing mostly between the cells.

Establishment of Infection

- Infection is the process by which pathogens establish contact with susceptible host and procure nutrients from them.
- Following infection, pathogens grow, multiply, or both within the plant tissues and invade and colonize the plant to a lesser or greater extent.
- During infection, some pathogens obtain nutrients from living cells, often without killing the cells or others kill cells and utilize their contents as they invade them.
- During infection, pathogens release enzymes, toxins, and growth regulators etc that may affect the structural integrity of the host cells or their physiological processes.
- In response, the host reacts with a variety of defense mechanisms, which may result in varying degrees of protection of the plant from the pathogen.
- **Condition for a successful infection to occur are:**
  - Plant variety must be susceptible to the particular pathogen and at a susceptible stage.
  - Pathogen must be in a pathogenic stage that can infect immediately without requiring a resting period.
  - Temperature and moisture conditions in the environment of the plant must favor the growth and multiplication of the pathogen.
  - When these conditions occur at an optimum, the pathogen can invade the host plant up to the maximum of its potential, even in the presence of plant defenses, and, as a consequence, disease develops.

Localized Vs Systemic Infection

Localized Infection

- Many infections caused by fungi, bacteria, nematodes, viruses, and parasitic higher plants are local.
- They involve a single cell, a few cells, or a small area of the plant.
- These infections may remain localized throughout the growing season or they may enlarge slightly or very slowly.

Systemic Infection

- Pathogen from one initial point in a plant, spreads and invades most or all susceptible cells and tissues throughout the plant.
- Infections enlarge more or less rapidly and may involve an entire plant organ, a large part of the plant (a branch), or the entire plant.
- Infections caused by fastidious xylem- or phloem inhabiting bacteria, mollicutes, and protozoa and natural infections caused by viruses and viroids are systemic.
- Some downy mildew pathogens and some fungi causing smuts and rusts, also invade their hosts systemically, the older mycelium degenerates and disappears and only the younger mycelium survives in actively growing plant tissues.
- Vascular wilt fungi and bacteria invade xylem vessels internally, but they are usually confined to a few vessels in the roots, the stem, or the top of infected plants; only in the final stages of the disease do they invade most or all xylem vessels of the plant.

Invasion

Bacteria

- Invade tissues intercellulary, although when parts of the cell walls dissolve, bacteria also grow intracellularly.
- Bacteria causing vascular wilts also invade the xylem vessels.

Viruses and viroids

- Invade tissues by moving from cell to cell intracellularly.
- Invade all types of living plant cells.

Mollicutes and protozoa

- Invade phloem sieve tubes and perhaps a few adjacent phloem parenchyma cells.

Fastidious bacteria

- Invade xylem vessels and a few invade only phloem sieve tubes.

Nematodes

- Most nematodes invade tissues intercellularly, but some can invade intracellularly as well.
- Many nematodes do not invade cells or tissues at all but feed by piercing epidermal cells with their stylets.
- Spread of a pathogen into the host is called invasion.
- Different pathogens *viz.*, fungi, bacteria, nematodes, viruses, viroids, mollicutes, fastidious bacteria, and protozoa invade hosts in different ways and to different extents.

Fungi

- Most fungi spread into all the tissues of the plant organs they infect, either by growing between the cells as an intercellular mycelium or directly through the cells as an intracellular mycelium.
- Fungal pathogens causing apple scab and black spot of rose, produce mycelium that grows only in the area between the cuticle and the epidermis (subcuticular colonization) .
- Fungal pathogens causing powdery mildews, produce mycelium only on the surface of the plant but send haustoria into the epidermal cells.
- Fungal pathogens that causes vascular wilts invade the xylem vessels of plants.

Primary Determinant *vs* Secondary Determinant

Primary determinant

- The toxin required for pathogenicity is called primary determinant.
- It is essential for colonization by the producing pathogens and disease development in the host.
- The primary determinant is otherwise called pathogenicity factor.

Secondary determinant

- The toxin required for virulence is called secondary determinant.
- Secondary determinant accounts for certain symptoms or contribute to virulence, but are not essential for colonization.
- The secondary determinant is otherwise called virulence factor.

3

Enzymes and Toxins in Relation to Plant Disease

Enzymes and Toxins

- Enzymes and toxins are probably more important in plant disease development.
- Pathogens produce these substances either in the normal course of their activities (constitutively) or when they grow on certain substrates such as their host plants (inducible).
- **Plant pathogenic enzymes**
  - Disintegrate the structural components of host cells.
  - Break down inert food substances in the cell.
  - Affect components of its membranes and the protoplast directly.
  - Thereby interfering with its functioning systems.
- **Toxins seem to**
  - Act directly on protoplast components and interfere with the permeability of its membranes and with its function.

Enzymes in Plant Disease

- Enzymes are generally large protein molecules.
- They catalyze organic reactions in living cells and in solutions.
- Each enzyme, being a protein, is coded for by a specific gene.
- Some enzymes are present in cells at all times (constitutive).
- Many are produced only when they are needed by the cell in response to internal or external gene activators (induced).
- Each type of enzyme often exists in several forms known as isozymes.
- Isozymes carry out the same function but may vary from one another in several properties, requirements, and mechanism of action.
- Because most kinds of chemical reaction that occur in a cell are enzymatic, there are almost as many kinds of enzymes as there are chemical reactions.

Enzymatic Degradation of Cell Wall Substances

- Aerial plant part surfaces consist primarily of cuticle and/or cellulose, whereas root cell wall surfaces consist only of cellulose.
- Epidermal cell walls may also contain suberin and lignin.
- The degradation of each of these substances is brought about by the action of one or more sets of enzymes secreted by the pathogen.

Cuticular Wax

- Plant waxes are found as continuous layers outside or within the cuticle of many aerial plant parts.
- Pathogens, e.g., *Puccinia hordei*, produce enzymes that can degrade waxes.

Cutin

- Cutin is the main component of the cuticle.
- Many fungi and a few bacteria produce cutinases and/or nonspecific esterases, i.e., enzymes that can degrade cutin.
- The fungus *Botrytis cinerea*, the cause of numerous types of diseases on many plants, produces a cutinase and a lipase, both of which break down cutin.

Pectic Substances

- Pectic substances constitute the main components of the middle lamella.
- Several enzymes degrade pectic substances and are known as pectinases or pectolytic enzymes.
- Bacterial wilt of solanaceous crops caused by *Ralstonia solanacearum*, pectinolytic enzymes collectively are essential for disease to develop by accelerated colonization and enhanced aggressiveness by bacteria.
- *Colletotrichum*-caused anthracnoses, the fungus produces one pectin lyase that is a key virulence factor in disease development.
- Pectin–degrading enzymes are produced and play a role in the ability of nematodes, such as the root knot nematode, *Meloidogyne javanica*, for the penetration of root tissues.
- Pectic enzymes are produced by germinating spores and, apparently, acting together with other pathogen enzymes (cutinases and cellulases), assist in the penetration of the host.

Cellulose

- Cellulose-degrading enzymes (cellulases) produced by several phytopathogenic fungi, bacteria, and nematodes and by parasitic higher plants.

Function of cellulase enzyme

- Play a role in the softening and disintegration of cell wall material in living plant tissues.
- Facilitate the penetration and spread of the pathogen in the host .
- Further participate indirectly in disease development by releasing, from cellulose chains, soluble sugars that serve as food for the pathogen and, in the vascular diseases, by liberating into the transpiration stream large molecules from cellulose, which interfere with the normal movement of water.
- In the bacterial wilt of tomato, production of an endocellulase by the bacterium was required for the latter to be pathogenic and induce the disease.

Pectic Substances

- Pectic substances constitute the main components of the middle lamella.
- Several enzymes degrade pectic substances and are known as pectinases or pectolytic enzymes.
- Pectin methyl esterases, remove small branches off the pectin chains.
- Pectinases, called polygalacturonases, split the pectic chain by adding a molecule of water and breaking (hydrolyzing) the linkage between two galacturonan molecules.
- Pectinases, called Pectin lyases, split the chain by removing a molecule of water from the linkage, thereby breaking it and releasing products with an unsaturated double bond.
- Polygalacturonases and pectin lyases occur in types that either can break the pectin chain at random sites (endopectinases) and release shorter chains, or can break only the terminal linkage (exopectinases) of the chain and release single units of galacturonan.

Pectin-degrading enzymes shown to be involved in the production of many fungal and bacterial diseases, particularly those characterized by the soft rotting of tissues.

Cellulose

- Cellulose is a polysaccharide.
- Cellulose-degrading enzymes (cellulases) produced by several phytopathogenic fungi, bacteria, and nematodes and by parasitic higher plants.
- Role of cellulolytic enzymes

- Play a role in the softening and disintegration of cell wall material in living plant tissues.
- Facilitate penetration and spread of the pathogen in the host.
- Cause collapse and disintegration of the cellular structure.
- Thereby aiding the pathogen in the production of disease.
- In the bacterial wilt of tomato, production of an endocellulase by the bacterium was required for the latter to be pathogenic and induce the disease.

Hemicelluloses *(Cross-Linking Glycans)*

- Complex mixtures of polysaccharide polymers.
- Hemicelluloses are a major constituent of the primary cell wall.
- Several hemicellulases seem to be produced by many plant pathogenic fungi.
- The particular enzymes are called xylanase, galactanase, glucanase, arabinase, mannose etc produced by pathogens, depending on the monomer released from the polymer on which they act.
- The nonenzymatic breakdown of hemicelluloses by activated oxygen, hydroxyl, and other radicals produced by attacking fungi also occurs.
- Fungal pathogens produce these enzymes and oxidative agents but it is still not clear how they contribute to cell wall breakdown or to the ability of the pathogen to cause disease.

Lignin

- Lignin is an amorphous, three-dimensional polymer.
- Most common basic structural unit of lignin is a phenylpropanoid.
- Most of the lignin in the world is degraded and utilized by a group of basidiomycetes called white rot fungi.
- White rot fungi secrete one or more enzymes (ligninases), which enable them to utilize lignin.
- Several ascomycetes and imperfect fungi and even some bacteria, apparently produce small amounts of lignin-degrading enzymes and cause soft rot cavities in wood they colonize.

Suberin

- Suberin is found in certain tissues of various underground organs, such as roots, tubers, and stolons, and in periderm layers, such as cork and bark tissues.
- Suberins are also formed in response to wounding and to pathogen-induced defenses of certain organs and cell types.

- Typical suberization occurs, for example, in cut potato tubers where browning and encrustation develop in the form of multilamellar areas consisting of alternating polyaliphatic and polyaromatic layers. These layers are impermeable and help strengthen the cell wall and limit water loss through the wound.
- Although plants obviously produce enzymes that synthesize suberin, it is not known whether or how pathogens break it down during infection.

Enzymatic Degradation of Substances Contained in Plant Cells –

- Most kinds of pathogens live all or part of their lives in association with or inside the living protoplast.
- These pathogens obviously derive nutrients from the protoplast.
- Some of the nutrients, e.g., sugars and amino acids, are molecules sufficiently small to be absorbed by the pathogen directly.
- Some of the other plant cell constituents, however, such as starch, proteins, and fats, can be utilized only after degradation by enzymes secreted by the pathogen.

Starch

- Starch is the main reserve polysaccharide found in plant cells.
- Starch is a glucose polymer and exists in two forms: amylose, and amylopectin.
- Most pathogens utilize starch, and other reserve polysaccharides, in their metabolic activities.
- The degradation of starch is brought about by the action of enzymes called amylases.
- The end product of starch breakdown is glucose and it is used by the pathogens directly.

Proteins

- Plant cells contain different proteins, which play diverse roles as catalysts of cellular reactions (enzymes) or as structural material (in membranes and cell walls).
- All pathogens seem to be capable of degrading many kinds of protein molecules.
- The plant pathogenic enzymes involved in protein degradation are proteases or proteinases or, occasionally, peptidases.

Lipids

- Various types of lipids (Oils and fats) occur in all plant cells.

- Most important being phospholipids and glycolipids.
- Both of which, along with protein, are the main constituents of all plant cell membranes.
- Several fungi, bacteria, and nematodes are known to be capable of degrading lipids.
- Lipolytic enzymes, called lipases, phospholipases etc. hydrolyze liberation of the fatty acids from the lipid molecule.
- The fatty acids are presumably utilized by the pathogen directly.

Cell wall and membrane degrading enzymes produced by certain plant pathogens

Enzymes	Substrate	Pathogen
Cutinases	Cutin	• *Fusarium solani* • *Colletotrichum gloeosporioides*
Cellulases	Cellulose	• *Erwinia carotovora* • *Ralstonia solanacearum* • *Aschochyta pisi* • *Rhizoctonia solani*
Hemicellulases	Hemicellulose	• *Sclerotium rolfsii* • *Erwinia carotovora* • *Penicillium expansum*
Proteinases	Protein	• *Rhizoctonia solani* • *Sclerotium rolfsii*
Ligninases	Lignin	• *Botrytis cinerea* • *Heterobasidium annosus*
Esterases	Suberin	• *Fusarium solani f sp pisi*

Microbial Toxins in Plant Disease

Definition

'Toxins' can be defined as low molecular weight, non enzymatic microbial products toxic to the higher plants.

General characteristics of toxins

- Toxins, however, are extremely poisonous substances and are effective in very low concentrations.
- Toxins act directly on living host protoplasts, seriously damaging or killing the cells of the plant.
- Many toxins exist in multiple forms that have different potency.
- Fungi and bacteria may produce toxins in infected plants as well as in culture medium.

- Toxins injure host cells either by affecting the permeability of the cell membrane or by inactivating or inhibiting enzymes and then interrupting the corresponding enzymatic reactions.
- Certain toxins act as antimetabolites and induce a deficiency for an essential growth factor.
- Toxin differs from other microbial products like 'mycotoxins' and 'antibiotics' which are toxic to the animals and microbes (except producers).

Classification of Toxins

Toxins are classified as

1. Phytotoxins
2. Vivotoxins
3. Pathotoxins

1. Phytotoxins

- Toxins produced by pathogens which are toxic to plants but not considered as of primary importance during pathogenesis, is called as phytotoxin.
- Non specific toxins are considered as phytotoxins.

2. Vivotoxins

- The toxin produced *in vivo* (in infected tissues) which function in disease development but not as the initial inciting agent, is termed as vivotoxin.

3. Pathotoxins

- Pathotoxins are the toxins which are major determinants of disease and of pathogenicity.
- Pathotoxin induces all the typical symptoms in reasonable concentration and is co-related with pathogenicity.
- Host specific toxin can be considered as pathotoxins.

Classification of toxins on the basis of host specificity

- Host non specific or Host-non selective toxin
- Host-Specific or Host-selective toxins

Host non specific or Host non selective toxin

- Toxins that affect a wide range of host plants.
- Toxic substances produced by phytopathogenic microorganisms shown to produce all or part of the disease syndrome not only on the host plant,

but also on other species of plants that are not normally attacked by the pathogen in nature.

- Such toxins, called nonhost-specific or nonhost-selective toxins.
- These toxins increase the severity of disease caused by a pathogen.
- They affect the virulence of the pathogen, but are not essential for the pathogen to cause disease.
- They do not determine the pathogenicity of the pathogen.
- It includes phytotoxin and vivotoxin.
- Ex. Tabtoxin Tentoxin, and Phaseolotoxin, Cercosporin

Host-Specific or Host-Selective Toxins

- A host-specific or host-selective toxin is a substance produced by a pathogenic microorganism that, at physiological concentrations, is toxic only to the hosts of that pathogen and shows little or no toxicity against nonsusceptible plants.
- Most host-specific toxins must be present for the producing microorganism to be able to cause disease.
- Host-specific toxins shown to be produced only by certain fungi (*Cochliobolus*, *Alternaria*, *Periconia*, *Phyllosticta*, *Corynespora*, and *Hypoxylon*), although certain bacterial polysaccharides from *Pseudomonas* and *Xanthomonas* have been reported to be host specific.
- It includes pathotoxins.
- **Examples:** T Toxin, *Victorin, or HV Toxin, HC Toxin*

Host Specific toxins			
Toxin	**Pathogen**	**Host**	**Chemical nature**
AK-toxin	*Alternaria kikuchiana*	Pear	Altenin (furanose ring)
HV- toxin	*Helminthosporium victorae*	Oats	Victoxinine
HC-toxin	*Helminthosporium carbonum*	Corn	Cyclic peptide
AM- toxin	*Alternaria mali*	Apple	Alternariolide
AB-toxin	*Alternria brassicae*	Brassica spp.	Destruxin B
PC-toxin	*Periconia circinata*	Sorghum	Proteinaceous
Host Non specific toxins			
Tentoxin	*Alternaria alternata*	Onion	Cyclic tetrapeptide
Tabtoxin	*Pseudomonas tabaci*	Tobacco	Aminoacid terivative
Alternatic acid	*Alternaria solani*	Potato	Hemiquinone derivative
Cerato-ulmin	*Ceratocystis ulmi*	Elm	Large M. carbohydrate
Pyricularin	*Pyricularia oryzae*	Rice	Nitrogen containing compound
Fusaric acid	*Fusarium spp.*		5-n-butypicolinic acid

Role of Toxins in Pathogenicity and Virulence

- Virulence is a quantitative term denoting relative disease inducing ability of a microorganism.
- Pathogenicity refers to the capacity of an organism to induce disease.
- A change in host range indicates a change in pathogenicity while a change in severity of disease indicates a change in virulence.
- But a question arises, when does pathogenesis begin and how much must the pathogen spread for colonization to be successful.
- Toxin is not required for penetration. Both toxins produced and no produced penetrate the host tissues.
- Initial colonization (upto24h) of both the isolates is also same. Only subsequent invasion is different.
- It is most likely that toxins play important role only as virulence factor rather than pathogenicity factor.
- In other words, toxin is important in necrotrophic phase rather than in biotrophic phase.
- All the toxins isolated so far are from saprophytic pathogens.
- No toxin has been isolated from obligate parasites like powdery mildews, downy mildews and rusts.

4

Philosophy of Defence in Plants

- Plants cannot move, they have to resist or tolerate the pathogen standing on the same site.
- Plants have developed very strong preformed defence system in the form of structural and toxic barriers.
- Plants cannot eliminate but can only limit (contain) the invaders.
- Plants can tolerate greater drain of nutrients. Often they sacrifice some of the tissues to restrict the pathogen.
- Division of labour is not that developed, hence, every cell has to perform functions including defense.
- Plant cells contain anti-microbial substances in compartments, which help the cell to defend it.
- Plant metabolism is influenced by environmental conditions, and in turn the defense mechanism is influenced.
- Vascular system in plants is not efficient, movement through apoplast is passive while loading in symplast is very selective (toxic molecules cannot be transported).
- Every individual plant has its own defence system. Nothing operates at population or community level.
- Primitive system akin to animal immunity system is likely to operate; common antigenicity in compatible host-pathogen interactions reported.

Mechanisms of Defense used by Plants

Two types

1. **Pre-existing or Passive defense (**Defense present on or inside the plant since beginning)

   a. Pre-existing structural defense

   b. Pre-existing biochemical defense

2. **Post infection or Active defense** (Defense in response to attack by the pathogen)

a. Post infection structural defense
b. Post infection biochemical defense

1. Pre existing or Passive defenses

- External or internal structural barriers existing before pathogen attack.
- Called preexisting defense structures or passive or static or anti infection structures.
- Initiating from periphery, the following barriers occurring in different host plants are described.
  - Wax and cuticle
  - Epidermal layer
  - Natural openings
  - Internal structures

Pre-infectional morphological and structural defense mechanisms against plant pathogens

Sl.No	Defense barrier	Pathogen	Host
1	**Waxes and Cuticle**	*Albugo candida*	Brussels sprout
		Puccinia graminis	Berberries
		Melampsora lini	Linseed
		Colletotrichum cooffeanum	Coffee
2	**Structure of natural opening**		
	Broad cuticular ridge projection over the stomata	*Xanthomonas campestris pv.citri*	Mandarins citrus
	Mechnical (Sclerenchymatous) tissues	*Entyloma oryzae*	Rice
	Late openining of stomata	*Puccinia graminis tritici*	Wheat
3	**Thick and tough outer cell wall** (Presence of silicic acid and lignified epidermal cells)	*Pyricularia oryzae*	Rice

Pre Existing Biochemical Defense

- When biochemical defense exist in the plant even before the attack of the pathogen.
- Plants possess certain metabolic processes in their cells which imparts resistance to them against pathogen attacks.
- This type of defense adopted by the plants is called "biochemical defense".

1. Release of antimicrobial compounds
2. Inhibitor present in plant cells:

3. Lack of essential factors
4. Recognition factors:
5. Essential nutrients and growth factors
6. Common antigen in host plant

Pre existing biochemical defense

1. Release of antimicrobial compounds

- Plant commonly exude organic substances from above ground parts and roots, containing amino acids, sugars, organic acids, enzymes, glycosides etc.
- Some of the compounds occurring in exudates are inhibitory.
- These inhibitory substances directly affect the micro organism or encourage certain group to dominant the environment and function as antagonists of the pathogen.
- **For instance,**
- Certain varieties of linseed resist infection of *F. oxysporum f sp lini* because of the presence of HCN in their root exudates.
  - Presence of HCN in roots of sorghum and maize is also reported which might be responsible for reduction of pigeonpea wilt caused by *Fusarium udum* in plots where sorghum is grown as mixed crop with pigeonpea.
  - Red scale of onion (red varieties) conatin protocatechuic acid and catechol, which may exude in drops and impart resistance to attack of *Colletotrichum circinanas* (onion smudge disease).
  - The root exudates favor development of plant growth promoting rhizobacteria (PGPR), which suppresses the plant pathogens and induces resistance.

2. Inhibitor present in plant cells

- Plant cells contain certain pre-existing inhibitory substances which mainly play a defensive role against the particular pathogen.
- Phenols and quinones are two classes of antimicrobial compounds produced by some plants
- Lactones, cyanogenic glucosides, saponins, terpenoids, stilbenes and tannins are plant-produced compounds associated with pathogen resistance.
- Saponins are a class of phytoanticipins that destroy membrane integrity in saponin-sensitive parasites, released following wounding or infection.

- Another group of defensive compounds are the plant defensins, which interfere with pathogen nutrition and retard their development.
- Secreted defensins can create an antimicrobial microenvironment for germinating seeds and accumulated defensins can provide defence against insect-transmitted viruses in flowers, leaves and tubers.
- There are also proteins, both constitutive and induced that play a role in plant defence.

3. Recognition factors

- First step in infection process is cell to cell communication between pathogens and host.
- Plant of species or varieties may not infected by a pathogen if there surface cells lack specific recognition factors.
- If the pathogen does not recognize the plant as one of its hosts it may not adhere to the host surface or it may not produce infection substances such as enzymes or structures like appresoria and haustoria.
- These recognition molecules are of various types of oligosaccharides, polysaccharides and glycoproteins.

4. Essential nutrients and growth factors

- Majority of the obligate parasites and some of the facultative saprophytes, particularly among the fungi, are usually host specific.
- They grow only on their specific variety of host.
- This happen because of host specialized pathogens need certain specific substances.
- These substances are only available in adequate quantities in the host they infect.
- Hence, absence of these specific substances makes the other varieties inappropriate host for such a pathogen.
- For instance, in seedling disease of several crops caused by *Rhizoctonia solani*, successful infection depends on formation of infection cushion from which infection peg develops and causes penetration.
- Formation of these cushions is induced by certain essential nutrients in the host.
- In resistant plants, lack of such nutrients results in resistance to *R. solani*.
- Sometimes, a particular race of pathogen loses its ability to synthesize a certain substance owing to mutation and this makes the pathogen race non specific.

5. Common antigen in host plant

- Presence of common antigen (protein) in both host and pathogen determines disease occurrence in the host.
- But, if, the antigen is absent in the host and present only in the pathogen, it makes the host resistant to that particular pathogen.
- For instance, the varieties of linseed which have an antigen common to their pathogen *Melampsora lini* prove to be susceptible to rust disease.
- In contrary, the linseed varieties not having the antigen in them but occurring in the pathogen are resistant to the pathogen.
- Similar is the case with leaf spot disease of cotton caused by *Xanthomonas compestris pv. malvacearum.*

Pre-infectional biochemical defense mechanisms against plant pathogens

Sl.No	Defense barrier	Pathogen	Host
1	**Inhibitors released by the plant in the environment**		
	Glucoside	Fusarium spp.	Flax
	Malic acid	*Ascochyta rabiei*	Gram
	Catechol and protocatechuic acid	*Colletotrichum circinans*	Onion
	Cutin acid	*Gloesporium limetticola*	Citrus
2	**Phenolic substances**		
	Polyphenols	*Venturia inaequalis*	Apple
		Pyricularia oryzae	Rice
		Venturia pirina	Pear
3	**Inhibitory substances present in the plant cells**		
	Chlorogenic acid	*Streptomyce scabies*	Potato
	Tuliposide	*Fusarium oxysporum f.sp. tulipae*	Tulip
	Avenacin	*Ophiobolus graminis*	Wheat
	Arbutin	*Erwinia amylovora*	Pear
4	**Absence of nutrients required by the pathogen**		
	Histidine	*Erwinia amylovora*	Potato
	Sugar	*Ceratocystis ulmi*	Elm
		Alternaria solani	Potato
		Cochliobolus sativus	Barley
	Choline	*Venturia inaualis*	Apple
	Myoinositol	*Rhizoctonia aurecina*	Plum

2. Post Infection or Active or Induced Defense

After the pathogen successfully penetrates the pre-existing defense barriers of the host, the latter, somehow, generally develops definite defense structures inside them in response to the reaction taking place between host and pathogen. These structures so developed are generally of four types:

- Induced histological defense
- Induced cellular defense
- Induced cytoplasmic defense
- Induced hypersensitive defense

Post-infectional morphological and structural defense mechanisms against plant pathogens

Sl.No	Defense barrier	Pathogen	Host
1	Formation of cork layer	*Rhizoctonia solani*	Potato
		Coccomyces prunophorae	Plum
	Formation of tyloses	*Fusarium oxysporum f sp batatas*	Sweet potato
		Verticillium- alboatrum	Tomato
3	Formation of abscission layer	*Xanthomonas arboricola pv. pruni*	Sour cherry
		Clsterosporium carpophilum	Peach
3	Sheathing of hyphae	*Phytophthora infestans*	Potato
4	Gum deposition	*Pyricularia oryzae*	Rice
		Drechslera oryzae	Rice
		Physalopora cydoniae	Apple

Defense through the Hypersensitive Response

- Hypersensitive response is considered a biochemical rather than a structural defense mechanism.
- When pathogen establishes contact with the cells of host, the nucleus moves toward the invading pathogen and soon disintegrates.
- At the same time, brown, resin-like granules form in the cytoplasm, first around the point of penetration of the pathogen and then throughout the cytoplasm.
- As the browning discoloration of the plant cell cytoplasm continues and death sets in, the invading hypha begins to degenerate.
- The hypha does not grow out of such cells in most cases, and further invasion is stopped.
- This type of necrotic defense is quite common, particularly in diseases caused by obligate fungal parasites and by viruses, bacteria, and nematodes.

- The necrotic tissue isolates the parasite from the living substance on which it depends for its nutrition and, thereby, results in its starvation and death.
- It signifies the concentration of numerous biochemical cell responses and antimicrobial substances that neutralize the pathogen.

Post Infection or Induced Biochemical Defense

- These mechanisms restrict the spread of the pathogen after infection is established and contain the damage to host tissues.
  1. Inactivation of Enzymes and Toxins
  2. Defense through Pathogenesis-related Proteins
  3. Defense through Phytoalexins
  4. Defense through Active Oxygen Species and Lipoxygenases

1. Inactivation of Enzymes and Toxins

- The role played by enzymes and toxins of pathogens during pathogenesis is well established.
- The hemibiotrophs and necrotrophs employ more of these substances for causing host tissue damage as compared to biotrophs.
- The defense strategy of resistant plants, through activity of phenols, tannins, proteins as enzymes inhibitors.
- the phenolics are not anti fungal but make pathogen ineffective by neutralizing their enzymes.
- In immature grape fruits catechol- tannin is known to inhibit enzymes produced by *Botrytis cinerea.*
- In resistant varieties of plants, which get diseased as a result of toxins produced by the pathogens, the pathogenic toxins released are detoxified by the ability of metabolic processes.

- **For example**
  - *Pyricularia oryzae*, while causing blast disease of rice, releases picolinic acid and pyricularin toxin inside the host. into other compounds.
  - Though susceptible varieties get affected by these toxic substances, the resistant varieties convert 40-60% of picolinic acid into its methylester and N-methylpicolinic acid, pyricularin into other compounds.
  - These resistant compounds are non toxic to rice plants.

Defense through detoxification of pathogen toxin

Detoxifying compound	Pathogen toxin	Disease	Pathogen
Chlorogenic acid	Piricularin	Blast of rice	*Pyricularia oryzae*
Dehydroascorbic acid	AK toxin	Black spot of pear	*Alternaria kikuchiana*
High concentration of cysteine	Fusaric acid	Wilt of cotton	*Fusarium oxysporum* f sp *vasinfectum*
		Wilt of tomato	*Fusarium oxysporum* f sp *lycopersici*

2. Defense Through Pathogenesis-related Proteins

- Called PR proteins.
- A structurally diverse group of plant proteins that are toxic to invading fungal pathogens.
- Widely distributed in plants in trace amounts.
- Produced in much greater concentration following pathogen attack or stress.
- PR proteins exist in plant cells intracellularly and also in the intercellular spaces (cell walls).
- Different plant organs, e.g., leaves, seeds, and roots, may produce different sets of PR proteins.
- PR proteins are either extremely acidic or extremely basic and therefore are highly soluble and reactive.
- At least 14 families of PR proteins are recognized.
- Better known PR proteins are
  - PR1 proteins (antioomycete and antifungal)
  - PR2 (b-1,3-glucanases)
  - PR3 (chitinases)
  - PR4 proteins (antifungal)
  - PR6 (proteinase inhibitors), thaumatine-like proteins, defensins, thionins, lysozymes, osmotinlike proteins, lipoxygenases, cysteine-rich proteins, glycine-rich proteins, proteinases, chitosanases, and peroxidases.
- Healthy plants attack by pathogens, treatment with elicitors, wounding, or stress induce transcription of a battery of genes that code for PR proteins.
- The signal compounds responsible for induction of PR proteins include salicylic acid, ethylene, xylanase, the polypeptide systemin, jasmonic acid, and probably others.

Role of Pathogenesis-related proteins

- Antifungal and other antimicrobial activity.
- Inhibit spore release and germination.
- Strengthening of the host cell wall and its outgrowths and papillae.

Families of pathogenesis –related Proteins (PRPs)

Sl. No	Protein family	Protein activity	Targeted pathogen sites
1	PRP-1	Unknown	Active against oomycetes
2	PRP-2	B-1,3 glucanase	Cell wall glucan of fungi
3	PRP-3	Endochitinase	Cell wall chitin of fungi
4	PRP-4	Endochitinase	Cell wall chitin of fungi
5	PRP-5	Taumartin-like	Active against oomycetes
6	PRP-6	Proteinase inhibitor	Active on nematodes and insects
7	PRP-7	Endoroteinase	Microbial cell wall dissolution
8	PRP-8	Endochitinase with lysozyme activity	Cell wall chitin of fungi, and mucopeptide wall of bacteria
9	PRP-9	Peroxidase	
10	PRP-10	Ribonuclease activity	Viral RNA
11	PRP-11	Endochitinase	Cell wall chitin of fungi
12	PRP-12	Defensin	Antifungal and antibacterial activity
13	PRP-13	Thionin	Antifungal and antibacterial activity
14	PRP-14	Lipid transfer proteins	Antifungal and antibacterial activity
15	PRP-15	Oxalate-oxidase	Produces H_2O_2 that inhibits microbes and also stimulates host defense
16	PRP-16	Oxalate-oxidase-like with suuperdismutase	Produces H_2O_2 that inhibits microbes and also stimulates host defense

Examples

- Some of the PR proteins, e.g., b-1,3-glucanase and chitinase, diffuse toward and affect (break down) the chitin-supported structure of the cell walls of several plant pathogens.
- Lysozymes degrade the glucosamine and muramic acid components of bacterial cell walls.
- Lipoxygenases and lipid peroxidases generate antimicrobial metabolites as well as secondary signal molecules such as jasmonic acid.
- Plant defensins (cysteine-rich peptides) have antimicrobial activity and accumulate through the ethylene and jasmonic acid pathway.
- Plants genetically engineered to express chitinase genes show good resistance against the soilborne fungus *Rhizoctonia solani*.

- Tobacco plants treated with lipopolysaccharides obtained from the outer wall of gram-negative bacteria produced several PR proteins and exhibited enhanced defense responses in tobacco against *Phytophthora nicotianae*, including the production of a systemic response in the leaves of plants inoculated through the roots.

3. Defense Through Phytoalexins

- Toxic antimicrobial substances produced in appreciable amounts in plants only after stimulation by various types of plant pathogenic microorganisms or by chemical and mechanical injury.
- They are produced by healthy cells adjacent to localized damaged and necrotic cells in response to materials diffusing from the damaged cells.
- Phytoalexins are not produced during compatible biotrophic infections.
- They accumulate around both resistant and susceptible necrotic tissues.
- Resistance occurs when one or more phytoalexins reach a concentration sufficient to restrict pathogen development.
- Most known phytoalexins are toxic to and inhibit the growth of fungi pathogenic to plants.
- Some phytoalexins are also toxic to bacteria, nematodes, and other organisms.
- In most legumes, phytoalexins are isoflavonoids, and in the Solanaceae they are terpenoids.
- Phytoalexins include phaseollin in bean; pisatin in pea; glyceollin in soybean, alfalfa, and clover; rishitin in potato; gossypol in cotton; and capsidiol in pepper.
- Most phytoalexin elicitors are generally constituents of the fungal cell wall, such as glucans, chitosan, glycoproteins, and polysaccharides.
- The elicitor molecules are released from the fungal cell wall, by host plant enzymes.
- Most such elicitors are nonspecific, i.e., they are present in both compatible and incompatible races of the pathogen and induce phytoalexin accumulation irrespective of the plant cultivar.
- Most phytoalexin elicitors are of pathogen origin.
- Some elicitors, e.g., oligomers of galacturonic acid, are produced by plant cells in response to infection or are released from plant cell walls after their partial breakdown by cell walldegrading enzymes of the pathogen.

Some important phytoalexins

Sl. No	Phytoalexins	Producer Plant	Chemical nature
1	Resveratrol	*Vitis viifera*	Stilbene
2	Trifolirhazin	*Trifolium pratens*	Isoflavonoid
3	Isocoumarin	*Daucus carota*	Isocoumarin
4	Orchinol	*Orchis militaris*	Dihydrophenanthrene
5	Isomeamarone	*Ipomea batata*	Sesquiterpene
6	Rishitin	*Solanum tuberosum*	Sesuiterpene
7	Lubimin	*Solanum tuberosum*	Sesuiterpene
8	Camelexins	*Aradiopsis thaliana* *Camalenia sativa*	Thiazoyl substituted indole camalexin
9	Brassinins	*Brassica rapa*	Indole ring linked to a dithiocarbamate
10	Pisatin	*Pisum sativum*	Isofalvonoid
11	Maackiain	*Cicer arietinum*	Isofalvonoid
12	Phaseollidin	*Phaseolus vulgaris*	Isofalvonoid

Defense through Production of Secondary Metabolites: Phenolics

- Phenolics are the common compounds found in plants and possess an aromatic ring bearing a hydroxyl substituent.
- Phenolics are toxic to fungi, bacteria and viruses.
- Common phenolic compounds that are toxic to pathogens are produced and accumulate at a faster rate after infection.
- It happens especially in a resistant variety of plant relative to a susceptible variety.
- Examples of such phenolic compounds; chlorogenic acid, caffeic acid, and ferulic acid.
- In date palm tree roots, cell wall-bound hydroxybenzoic acid and sinapic acid increased 11–12 times as much in cultivars resistant to Fusarium than they did in susceptible cultivars.
- In peach, chlorogenic acid is present in quite high concentration both in immature fruit and in fruit of varieties resistant to the brown rot disease caused by the fungus *Monilinia fructicola.*

Role of Phenol Compounds

They inhibit

- Mycelia growth
- Spore germination
- Producion of cellulotic and pectic enxymes
- Activity of cellulotic and pectic enzymes and
- Production of toxins by the fungi

Phenolic substances	Crop
Chlorogenic acid	Potato
Catechol and Pyrocatechic acid	Onion
Avenacin	Oats
Hordatine	Barley
Tomatine	Tomato
Tuliposides	Tulip
Catechin and Isoquercitin	Cotton
Umbelliferone and Scopoletin	Sweet potato

4. Defense through Active Oxygen Species and Lipoxygenases

- In many host–fungus interactions, one of the first events detected in attacked host cells, or cells treated artificially with fungal elicitors, is the rapid and transient generation of activated oxygen species, including superoxide (O2-), hydrogen peroxide (H_2O_2), and hydroxyl radical (OH).
- Generation of superoxide and of other reactive oxygen species as defense response happens most dramatically in localized infections.
- However, it also occurs in general and systemic infections, as well as in plants treated with chemicals that induce systemic acquired resistance.
- These highly reactive oxygen species are released by the multisubunit NADPH oxidase enzyme complex of the host cell plasma membrane.
- They appear to be released in affected cells within seconds or minutes from contact of the cell with the fungus or its elicitors and reach a maximum activity within minutes to a few hours.
- Activated oxygen species trigger the hydroperoxidation of membrane phospholipids, producing mixtures of lipid hydroperoxides.
- Latter are toxic, their production disrupts the plant cell membranes, and they seem to be involved in normal or HR-induced cell collapse and death.
- Active oxygen species may also be involved in host defense reactions through the oxidation of phenolic compounds into more toxic quinones and into lignin-like compounds.
- The presence of active oxygen species, however, also affects the membranes and the cells of the advancing pathogen either directly or indirectly through the hypersensitive response of the host cell.

Cell wall-mediated Resistance in Plants

- Forms the first line of defense against pathogens.
- There exist three different forms of protection:

1. Inhibition of pathogen-derived cell wall-degrading enzymes;
2. Structural and chemical remodelling of the cell wall at sites of attempted penetration

- Killing of potential intruders by antimicrobial means.
- Structural and chemical defence is brought by.
- Cell wall appositions.
- Papillae formation with accumulation of lignin.
- Callose.
- Phenyl propanoids.
- Pathogen-related proteins.
- Reactive oxygen species.
- Cross-linked hydroxyproline-rich glycoproteins (HRGPs).

Hydroxyproline-rich Glycoproteins (HRGPs)

- Cell wall-associated defenses include effective transport and secretion of defence compounds in the direction of polarization depending on the elicitor perception.
- Cross-linking of HRGPs in cell walls provides added resistance against pathogen-derived cell wall-degrading enzymes.

5

Nature of Resistance to Disease

- Resistance coordinated by a number of internal and external factor.
- They operated to reduce the chance and degree of infection in host.
- First step in infection: Recognition of host by the pathogen and vice versa.
- Absence of recognition factor in the host could help it avoid infection by a particular pathogen.
- Heritable characters that contributes to resistance of the plant to disease:
    - Localization and isolation of the pathogen at the point of entry.
    - Reduction of harmful effect of toxic substance produced by the pathogen.
    - Inhibition of reproduction and further spread of the pathogen.

Results

- Pathogen is localized after varrying degree of invasion and colonization of host tissues.
- Indeed, there are only few diseases in which the pathogen is allowed to advance unchecked throughout the plant and to kill the entire plant.
- Any heritable character that enables a particular variety to complete life cycle under conditions that do not favor the development of the pathogen also contributes to resistance (DISEASE ESCAPE).

Contribution of the genes conditioning resistance in the host

- Gene provides genetic potential in plant for development of one or more morphological or physiological characters that contribute to disease resistance.
- Production on an inducible enzyme or fungitoxic substance needed for defense called stimulant (elicitor).
- Elicitors are secreted by the pathogen, reacts with a receptor molecule of a host cell., then transmits signals to other host cell molecules, activating plant defenses.

- On the contrary, if a pathogen mutant appears which does not secrete the particular elicitor, the pathogen infect the host without opposition and so cause disease and resistance is said to be broken down.
- Other possible ways through which could breakdown the resistance of host are through mutation in the pathogen that enable it to produce a substance which can react with and neutralize the defense toxin substance of host that is directed against the pathogens, and through mutation in the pathogen that would eliminate or block its own receptor cite on which the host defensive substance becomes attacked.

Genetics of Resistance through Hypersentsitive Response (HR)

- HR is a localized self induced cell death at the site of infection of a host plant in response to infection by pathogen in a particular resistant variety.
- Plant species as a whole may be host of pathogen but individual varieties of plant may be susceptible or resistant to a particular race of pathogen.
- Resistance through HR is result of gene for gene systems in which an avirulence gene (avr) gene in the pathogen corresponds to a resistance (R) gene in the host plant.
- HR occurs in disease caused by obligate and facultative pathogens virus, mollicutes, bacteria, fungi and nematodes.
- Mechanism
    - Recognition by the plant to specific signal molecules called elicitors, produced by the avirulence genes of the pathogen and recognized by R gene coded specific receptor molecules in the plant.
    - Such recognition causes activation of a cascade of host genes which then leads to hypersensitive response, inhibition of pathogen growth, thereby resistance.
    - It also leads to activation of numerous other defense related genes that result in other types of resistance, including HR & SAR.

Gene Coded Proteins

Avr protein: Protein coded by pathogen avr genes.

Characteristics

- Hydrophilic i.e water soluble.
- Lack signal sequences (streches of aminoacid)that allow protein to be secreted in the external medium).

- Avr gene protein secreted through membrane pore formed by proteins coded for HR and pathogenicity (hrp) genes, known as Hrp proteins (harpins).
- Avr protein may act as elicitors which react directly or indirectly with the product of the corresponding plant resistance R gene.

Structures

- Best known avr gene group is the Xanthomonas avr gene family, called ptb (for pathogenicity) genes by some.
- Members of this gene found in Xanthomonas bacteria
- Avr/ptb genes causes the HR and also required for induction of angular leaf spot symptom of cotton and for citrus canker.

Function of avr Gene proteins

- AvrD gene found in bacterium *Pseudomonas syringae pv glycinea.*
- AvrCf9 and avrCf4 found in tomato leaf mold fungus *Cladosporium fulvum.*
- Avr gene products of these pathogens are precoursor of elicitor peptides that are exported and the elicit HR in host carrying corresponding R gene for resistance.
- Elicitor resides on coat protein of TMV causing hypersensitive response in *Nicotiana sylvestris* tobacco carrying N gene for resistance.
- In some viral infections, viral coat protein produced within the cell appears to functions as specific elicitors that activates HR in host that carrying R gene for that virus.

Role of avr Genes in Pathogenicity and Virulence

- Most avr genes play no role in pathogenicity or virulence of the pathogen.
- However, some avr genes, for example, the avr Bs2 gene for bacterium *X. campestris* pv. *vesicatoria*, encode proteins that are also necessary for pathogenicity.
- Avr genes like pthA gene from *X.citri* and avrb6 for *X.campestris pv. malvacearum*, encodes protein that act as pathogencity or virulence factors. They enhance the virulence of a weakly pathogenic leaf spotting strain of *X. citrumelo.*
- Pth A is required for pathogenicity of *X.citri* on citrus to cause typical citrus canker disease on host and as avirulence genes , for example by causing pth A transofrmed strains of *X.phaseoli* and *X.campestris* pv. *malvacearum* that, infect beans or cotton, but not citrus, to cause HR on their respective hosts bean and cotton while remaining non-pathogenic on citrus.

hrp Genes

- Called Hypersensitive response and pathogenicity genes in bacteria.
- Mostly found in gram negative bacteria.
- Additional bacterial gene s seems to be able to cause
  - Visible disease on plant host.
  - To induce a hypersensitive response on certain plants that are not normally infected by bacteria.
  - To enable bacteria to multiply and reach high number in susceptible hosts.
  - Several hrp gene coded protein are called harpins, localized in the bacterial cell membrane.

Other Genes Involved in Pathogenesis and Virulence by Pathogen

- Pathogenicity factors encoded by pathogenicity genes (pat) and disease specific genes (dsp) are those involved in steps crucial for the establishment of disease.
- Such genes include those essential for:
  - Attachment of the pathogen to the plant surface.
  - Germination and formation of infection structures on the plant surface.
  - Penetration of the host.
  - Colonization of the host tissues.

6

Gene for Gene Hypothesis

1. Hypothesis first time proposed by HH Flor in Flax rust (*Melampsora lini*).
2. Later on operate in many other rusts, powdery mildews, apple scab, late blight of potato & other disease caused by fungi, several bacteria, viruses, parasitic higher plants, nematodes.

Concept

- Co-existance of host plants & their pathogens side by side in nature indicates that the two have been evolved together.
- Change in the virulence of the pathogens appear to be continuously balanced by changes in the resistance of the host, & vice versa.
- In that way a dynamic equilibrium of resistance and virulence is maintained, and both host and pathogen survive over considerable periods of time.
- Stepwise evolution of virulence and resistance can be explained by the gene for gene concept.

Hypothesis

Each gene that confers virulence to the pathogen there is a corresponding gene in the host that confer resistance to the host and vice versa.

Gene combination and disease reaction types in a Host - Pathogen system in which Gene for Gene concept for one gene operates

Generaly in host

1. Genes for resistance are dominant –R
2. Gene for susceptibility are recessive-r

In pathogen

1. Gene for avirulence (inability to infect) are dominant-A
2. Gene for virulence are recessive-a

	Resistance or susceptibility genes in plant		Minus sign indicate incompatible (resistance) reaction & no infection
Virulence or avirulence genes in the pathogen	**R (Resistant) Dominant**	**r (Susceptible) Recessive**	
A (Avirulent) dominant	AR (-)	Ar (+)	Plus sign indicate compatible (susceptible) reaction & infection develops
a (Virulent) recessive	aR (+)	ar (+)	

3. Outcome

Each gene in the host can be identified only by its counterpart gene in the pathogen and vice versa

Gene combination	Reaction	Reason
AR	Resistant	Host R gene recognizes avirulence gene (A)
Ar	Susceptible	Host lack gene for resistant (r) and so pathogen attack it with other gene of virulence
aR	Susceptible	Although host has gene for resistance, pathogen lack gene for avirulence
ar	Susceptible	Plant has no resistance (r) & pathogen is virulent (a)

Complementary interaction of two host genes for resistance and the corresponding two pathogen genes for virulence and their reactions types

Virulence (a) or avirulence (A) genes in the pathogen	Resistance (R) or susceptibility (r) genes in the plant				
		R1R2	**R1r2**	**r1R2**	**r1r2**
	A1A2	-	-	-	+
	A1a2	-	-	+	+
	a1A2	-	+	-	+
	a1a2	+	+	+	+

1	**Susceptible plant lacking gene for resistance are attacked by all race of the pathogen**
2	Pathogen races or individaul designated (a1a2) , which lack gene s for avirulence (a1a2) for each gene for resistnace of the host (**R1R2**) that infect all plant the have any combination of these genes since the a1a2 pathogen produces no elicitor molecule capable of triggering the host defense.
3	When a pathogen has one of the two genes for virulence (a1 or a2) , that is , it lacks one of the two genes of avirulence (A1 orA2) , then it can infect plants that have corresponding gene for resistance (**R1** or **R2**). Ex. Pathogen with A1a2 genes infects plant with **r1R2** but not those with **R1r2,** because **R1** can reognize the avr gene A1 and triggers defenses against it.

7

Management of Pathogens through Satellite, Antisense - RNA Ribozymes, Coat Protein, RNA Interference, Plantibodies, Hypovirulence, Cross Protection

Disease Control by Immunization

1. Most common improvement of host resistance to almost any pathogen is brought about by improving the genetic resistance of the host, that is by breeding and using resistance varieties.
2. Genetic engineering technology made possible the isolation of individual resistance (R) genes from resistant plants and the transfer of such genes into susceptible plants in which they induce the hypersensitive resistant response.
3. Thus, Genetic Engineering Technology combined with Conventional Plant Breeding, will provide one of the most effective tools for controlling plant diseases.
4. Inoculation of plants with certain pathogens often leads to temporary or nearly permanent immunization of the plants, that is, induced plant resistance to a pathogen to which the plants are normally susceptible.
5. Some of these treatments involve only viruses and are known as cross protection; others may involve different kind of pathogens and are known as induced or systemic acquired resistance.
6. Inoculation of plants with certain pathogens often leads to temporary or nearly permanent immunization of the plants, that is, induced plant resistance to a pathogen to which the plants are normally susceptible.

Cross Protection

Protection of a plant by a mild strain of viruses from infection by a strain of the same viruses that causes more severe symptoms.

Application

- Controlling viral diseases
    - Tomatoes with mild strains of Tobacco mosaic virus.
    - Citrus with mild strains of citrus Tristeza virus.
    - Papaya with mild strains of Papaya ring spot virus.

Limited Use in Crop Protection

Cross protection has not gained wide spread use due to

1. Non availability of appropriate mild strains of viruses.
2. Mild strains are not effective against all severe strains present in different localities.
3. Method for field crops is laborious.
4. Danger of mutation toward new more virulent strains and spread to other crops.
5. Crop protection ability of mild strain is lost after few years.

Defense through Plantibodies

1. Plants, do not have an immune system like that of humans and animals, that is, they do not produce antibodies.
2. Of late, transgenic plants have been produced that have been genetically engineered to incorporate in their genome, and to express, foreign genes, such as mouse genes that produce antibodies against certain pathogens.
3. Such antibodies, encoded by animal genes but produced in and by the plant, are called plantibodies.
4. Transgenic plants producing plantibodies against coat proteins of viruses, For example, artichoke mottle crinkle virus, to which they are susceptible, can defend themselves and show some resistance to infection by these viruses.
5. This type of plant immunization will yield great dividends by expressing in plants animal antibody genes that will produce antibodies directed against specific essential proteins of the pathogen viz., viral replicases or movement proteins and fungal and bacterial enzymes of attack.

Induced Resistance

Inoculation of plants with certain pathogens often leads to temporary or nearly permanent immunization of the plants, that is, induced plant resistance to a pathogen to which the plants are normally susceptible.

Resistant to pathogens could be induced in their hosts by

1. Treating the host with naturally occurring compounds obtained from the pathogen.
2. Treating hosts with unrelated natural compounds.

1. Treating the host with naturally occurring compounds obtained from the pathogen

- Coat protein of TMV.
- Proteinaceous component or glycoprotein fraction from a bacterium (*Pseudomonas solanacearum*).
- Lipid component from a fungus (*Phytophthora infestans*).
- Polysaccharide such as chitosan from a fungus.

2. Treating hosts with unrelated natural compounds

- Water soluble fraction of non pathogenic bacterium.
- Polysaccharide from a nonpathogenic fungus.
- Proteinaceous compounds isolated from an unrelated plant.

Examples

1. Bean and Sugarbeet inoculated with virus exhibit a greater resistance to infection by certain obligate fungal pathogens causing rusts and powdery mildews than do virus free plants.
2. In tobacco, TMV induces a systemic resistance not only to itself but also to unrelated viruses, to fungi like *Phytophthora nicotianae*, to bacteria like *Pseudomonas tabaci* and even to certain aphids.
3. Inoculation of tobacco with fungi *viz.*, *Chalara elegans* (*Thielaviopsis basicola*) or a bacterium *viz.*, *Pseudomonas syringae* induces systemic resistance to TMV.
4. Inoculation of apple with nonpathogenic bacterium induces resistance against fire blight of apple.

Plant Defense Inducers

1. Synthetic compounds which can induce resistance in plants to several viruses, fungi and bacteria are called plant defense inducers.
2. They includes salicylic acid and dichloroisonicotinic acid (INA).
3. They are applied by injection into the plant, spraying onto leaves or absorption through the petiole or through the roots.
4. Such synthetic chemicals reported as effective inducers of systemic acquired resistance against several pathogens like TMV, *Peronospora tabacina* and *Pseudomonas syringae.*

5. A number of work is going on in these area by public and private sector in developing compounds that would induce SAR to plant disease.
6. One such compound known as CGA 245704 that have no antimicrobial activity but mimic the biological induction of SAR in both dicot and monocot crops against many pathogens

Gene Editing

1. Gene editing is an exciting innovation that has captured the imagination of the research community, which allows scientists to make more targeted improvements within a plant's DNA to produce a disease and pest resistant crop.
2. It has the potential to solve real challenges for farmers and the planet, like reducing the need for pesticides and the use of energy, land, and water.
3. In agriculture, this process typically looks to improve a beneficial trait within an organism, or to remove an undesirable trait.
4. For years, "gene editing" was done through selective breeding in plants. But now we can make changes with more precision than ever before.
5. Gene editing tools, like CRISPR, are already helping researchers to make improvements within plant DNA.
6. These tools enable scientists to: deactivate unfavorable traits, such as disease vulnerability and support beneficial traits, such as drought tolerance or improved nutrition.
7. These tools have the potential to offer unmatched precision to farmers, allowing them to grow enough food while confidently reducing their use of natural resources.
8. By giving plants a better chance at survival—particularly in regions that struggle with hunger and malnutrition—gene editing can give people a better chance at life.

Management of pathogens through satellite RNA

Satellite RNA (satRNA)

1. Supernumerary RNA of some viruses.
2. Typically shares no nucleotide homology with its helper virus.
3. Depends on its helper virus for replication and encapsidation.
4. Presence of satRNA sometimes interferes with the replication and symptom expression of its helper virus.

5. Satellite RNA can have neutral, attenuated or intensified effects on the virulence of its helper virus owing to complicated interactions among satRNA variants, helper viruses and host genotypes.
6. Attenuated satRNAs used as biocontrol agents of plant diseases caused by their helper viruses.
7. Post-inoculation with satRNA could be a new way to control diseased plants pre-infected by helper viruses.
8. Could form a new therapeutic strategy to combat plant viruses using attenuation satRNAs.

Ways of biocontrol of viruses by satRNAs

Attenuated satRNAs have been used as **biocontrol agents** of plant diseases caused by their helper viruses in two ways.

1. By cross-protection
2. Creation of satRNA-transgenic plants.

1. Cross-protection

- Combination of satRNA and its helper virus is inoculated onto target plants in advance to prevent them from further pathogenicity of helper virus.

2. Creation of satRNA-transgenic plants

- $_{SAT}$RNA-transgenic plants created which convey phenotypes resistant against the virulence of the helper virus

Mechanisms by which satRNA triggers resistance to helper viruses

- Competition for viral replicase
- Gene silencing
- Possibly other immune actions of the host.

Methods of using satRNA as a biocontrol agent

1. Current methods of using satRNA as a biocontrol agent are preventive strategies.
2. They must be applied before infection of the target viruses.
3. Case study
    - Liu *et al.* (2018) identified a new satRNA isolate of CMV from tobacco in China and termed it satCMV TA-tb.
    - Isolate is 383 nt in length.
    - They found that co-inoculation with satCMV TA-tb and CMV_{Fny} remarkably inhibited symptoms caused by CMV.

Advantages of Post-inoculation with satRNA

1. Play an essential role in the control of diseased plants.
2. Simple and convenient to apply.
3. Partially compensate for the shortcomings of other strategies used to control diseased plants pre-infected with viruses.
4. Could also prevent new infections by its helper virus via cross-protection after post-inoculated satRNA interacts with pre-infected helper virus.

Satellite RNA, a feature that distinguishes plant viruses from animal viruses, is guaranteed to become a key player in the control of viral plant diseases.

Ribozyme

- Discovered by Sidney Altman and Thomas Cech.
- A ribozyme is a ribonucleic acid (RNA) enzyme that catalyzes a chemical reaction.
- The ribozyme catalyses specific reactions in a similar way to that of protein enzymes.
- Also called catalytic RNA.
- Ribozymes are found in the ribosome where they join amino acids together to form protein chains.
- Well-established natural ribozymes known to date are:
  - Hairpin
  - Hammerhead
  - Hepatitis delta virus (HDV)
  - Varkud satellite (VS)
  - And *glms* ribozyme
  - Group I and II introns
  - The ribosome
  - Spliceosome and RNase P, which are classified as large ribozymes.

Which type of RNA is present in ribozyme?

- RNase P RNA is the only characterized ribozyme that, unmodified, acts in trans on multiple substrates,
- Hence it is considered the only true, naturally occurring, RNA enzyme.

Examples of naturally occurring ribozymes

- RNase P.
- Peptidyl transferase 23S rRNA.

- GIR1 branching ribozyme.
- Leadzyme.
- Group I and Group II introns.
- Hairpin ribozyme.
- Hammerhead ribozyme.
- HDV ribozyme.

Functions of ribozymes

- Transfer RNA maturation
- Intron splicing
- Replication of RNA viruses or viroids
- The regulation of messenger RNA stability
- And protein synthesis.

Examples

- Ribozyme-mediated high resistance against potato spindle tuber viroid in transgenic potatoes.
- Ribozyme genes protecting transgenic melon plants against potyviruses.

Coat protein-mediated resistance

- Refer to the resistance caused by the expression of a virus coat protein (CP) gene in transgenic plants.
- Coat protein in transgenic plants can be used to enhance the production of movement proteins and coordinate the formation and size of virus replication complexes there by limiting the spread of cell to cell infection.

Coat protein-mediated resistance: the first example of pathogen-derived resistance for plants

- In 1985, Sanford and Johnston published the concept of parasite (or pathogen)-derived resistance (PDR).
- They suggested that the newly emerging genetic engineering technologies could be used to confer resistance by
    - Cloning the appropriate parasite gene
    - Modifying its expression if necessary
    - Transforming it into the host genome
- Then, when the corresponding gene product is expressed in the host, this could yield "faulty signals" which would disrupt the normal replication and infection cycle of the parasite.
- This disruption could then result in a resistance phenotype.

Why the concept was especially appealing as a resistance strategy for plant viruses?

This concept was especially appealing as a resistance strategy for plant viruses for several reasons.

1. First, all plant viruses are obligate intracellular parasites so this approach could be applied to any virus.
2. Second, most plant viruses have fewer than ~10 genes, so this strategy could be tested by experimenting with only a few genes.
3. And finally, the strategy could be applied without detailed knowledge of all the functions of a virus gene or protein.
4. It was simply hypothesized that inappropriate, unregulated expression of selected virus genes/proteins by the host plant would somehow lead to virus dysfunction and therefore resistance.

Genetically Engineered Tobacco Plants

- In 1986, Abel and colleagues reported the first example of genetically engineered virus resistance tobacco plant .
- They demonstrated resistance to TMV in transgenic tobacco plants engineered to express the TMV coat protein (CP) gene
- This resistance was demonstrated in two ways:
  - First, some of the CP+ plants showed a lower percentage of TMV-infected plants than did the nontransgenic controls.
  - Second, those that became infected developed symptoms later after inoculation than did the nontransgenic plants.
- By increasing the inoculum level, this delay in appearance of symptoms was reduced.
- None of the plants were immune to TMV challenge.

Examples of transgenic strategies resulting in enhanced disease resistance

Strategya	Gene/ construct	Donor	Recipient	Pathogen/disease	Phenotype	Reference
Part 1: Resistance against viruses						
PTGS (post translational gene silencing)	Full-length coat protein (CP) genes	Cucumber mosaic virus (CMV), zucchini yellow mosaicvirus (ZYMV), and watermelon mosaic virus (WMV)	Summer squash (*Cucurbita pepo* subsp. *ovifera* var. *ovifera*)	CMV (*Bromovirus*), ZYMV (*Potyvirus*), and WMV (*Potyvirus*)	Resistant, low virus titre	Reviewed by Fuchs and Gonsalves (2007)
PTGS	Full-length CP gene	Papaya ringspot virus (PRSV)	Papaya	PRSV (*Potyvirus*)	High levels of resistance	Reviewed by Fuchs and Gonsalves (2007)
PTGS	Full-length CP gene	Cowpea aphid-borne mosaic virus (CABMV) severe isolate SP	Yellow passion fruit (*Passiflora edulis* f. *flavicarpa*)	CABMV (*Potyvirus*)	Asymptomatic	Monteiro-Hara et al. (2011)
PTGS	Virus CP and other gene fragments, using CaMV35S promoter	Potato virus Y (PVY)	Potato (*Solanum tuberosum*)	PVY (*Potyvirus*)	No or negligible PVY infection	Bravo-Almonacid et al. (2012)
PTGS	CP gene including hairpin constructs using CaMV 35S promoter	Citrus tristeza virus (CTV)	Orange (*Citrus sinensis*) and grafted shoots	CTV (*Closterovirus*)	Low	Muniz et al. (2014)

RNA interference

- Also called Post transcriptional gene silencing PGTS.
- A common mechanism of PTGS is RNA interference (RNAi).
- Occurs after transcription (after synthesis of mRNA from DNA).
- Occurs due to destruction of m RNA.
- Occurs in both plants and animals.
- Provides resistance to viral infection.
- Prevents viral infection to the place of original virus attack.
- It is enhanced by transgenes, sequence homology etc.
- It is not related to DNA methylation.

Comparison of TGS and PTGS

Particulars		Transcriptional gene silencing (TGS)	Post Transcriptional gene silencing (PTGS)
1.	**Level of operation**	It occurs during synthesis of mRNA from DNA	It occurs during protein synthesis from mRNA
2.	**Cause**	It occurs due to DNA methylation	It occurs due to destruction of mRNA
3.	**Trangenes**	Enhance TGS	Enhance PTGS
4.	**Sequence homology**	Enhance TGS	Enhance PTGS
5.	**Control**	TGS events are meiotically heritable	PTGS events are not meiotically transmitted
6.	**Viral Infection**	Induces TGS	Induces PTGS
7.	**RNA Interference**	Not related to TGS	Related to PTGS
8.	**DNA methylation**	Related to TGS	Not related to PTGS
9.	**RNA degradation**	Not related to TGS	Related to PTGS

Mechanism of PTGS

- Post transcriptional gene silencing occurs in *nucleus* as well as *cytoplasm*.
- Main steps involved in PTGS in nucleus and cytoplasm are given below:

1. In Nucleus

- **Main steps**
  - Cleavage of mRNA into micro RNA (miRNA) precursors. It occurs by RNase III like enzyme Drosha.
  - Export/transport of miRNA precursor to cytoplasm. It takes place by means of export receptor exportin 5.
  - Processing of miRNA precursor into small interference RNA (siRNA) duplex. It takes place with the help of Dicer.

- Unwounding of siRNA duplex while assembling into miRNA/RISC.
- Binding of nature miRNA to Ago proteins resulting in degradation or cleavage (repression) of target mRNAs.

2. In Cytoplasm

- **Main steps**
  - Cleavage of double stranded RNA (ds RNA) into small interfering RNAs (si RNAs) of 21-25 nucleotides (nt). It is catalyzed by Dicer, an RNAse III type enzyme.
  - Then these siRNAs guide an RNA induced silencing complex (RISC) to destroy single stranded RNA (ssRNA).

Hypovirulence

- Reduced virulence of a pathogen strain as a result of the presence of transmissible double-stranded RNA.
- Antagonistic microorganisms may consist of avirulent strains of the same pathogen that destroy or inhibit the development of the pathogen.
- Hypovirulence in fungal plant pathogens refers to the reduced ability of selected isolates within a population of a pathogen to infect, colonize, kill, and (or) reproduce on susceptible host tissues and is often associated with fungal viruses and associated double-stranded RNA elements.

Biocontrol of chestnut blight with hypovirulent strains of the pathogen

- Hypovirulence is a symptom caused by a number of different fungal viruses and describes the reduced virulence of plant pathogenic fungi that are infected by these viruses.
- This symptom can simply be the result of
  - Debilitation of the fungus by viral infection,
  - or less commonly, can result from a perturbation of developmental processes of the fungus.
- The first described and characterized fungal virus responsible for hypovirulence was found to be responsible for transmissible hypovirulence of the chestnut blight fungus, *Cryphonectria parasitica*.
- This newly described virus became the type species of the family *Hypoviridae* and was given the name cryphonectria hypovirus 1 (CHV1).
- CHV1 is a double-stranded RNA virus that does not have a capsid and is not infectious.

- It appears to have evolved from the replicative form of a virus most closely related to the single-stranded RNA plus-sense plant potyviruses.
- CHV1 causes hypovirulence symptoms by perturbing fungal developmental processes, most likely as a result of its replication using *trans*-Golgi vesicle membranes.

Mycovirus as Biocontrol Agents

- The hypovirulence phenotype in the chestnut blight fungus (*Cryphonectria parasitica*) is an excellent and well-documented example for a mycoviral-induced phenotype that is currently being exploited for biological control.
- The debilitating disease of *Helminthosporium victoriae*, the causal agent of Victoria blight of oats, and the disease phenotype of the Dutch elm disease fungus *Ophiostoma novo-ulmi* are examples of pathogenic effects of dsRNA fungal viruses.

8

How do R Gene Confer Resistance?

Resistance (R) Gene of Plants

Plant varieties are resistance to certain pathogen races because they possess specific resistance gene (R) that enables the plant to remain resistant to pathogens carrying the corresponding avirulence (avr) genes.

Example

Hm1

1. First R gene isolated in year 1991.
2. Gene makes crop plants resistance to race 1 of fungus *Cochliobolus miyabeanus* (*Helminthosoprium carbonum*) causes leaf spot of corn.
3. Race 1of *Helminthosoprium carbonum* produces a host specific toxin HC-toxin which infect corn variety that lack Hm1 gene & are susceptible to fungus.
4. However, corn varieties resistant to race 1, expression of the Hm1 gene results in production of an enzyme called HC- toxin reductase and keep plant free from infection.

PTO gene

- Confer resistance in tomato to bacterial speck disease caused by *Pseudomons syringae* pv. *tomato*
- Pathogen carry avirulence gene avr-pto.
- Protein encoded by the PTO gene appears to be a serine threonine protein kinase, an enzyme suspected play role in HR.

Cf2, Cf4 &Cf9

- Confer resistance in tomato to leaf mold disease caused by *Cladosporium fulvum* races 2,4 & 9 that carry avirulence gene avr2, avr4 & avr9.

Tobacco N gene

- Confer resistance to TMV virus.

Flax L6 gene

- Confer resistance to rust fungus *Melampsora lini* race 6 carrying avr6 gene.

Xa21 gene

- Confer resistance to many races of BLB caused by *Xanthomonas oryzae.*

Classes of R Genes

Class	Function	Example of R gene
I	Encodes a detoxifying enzyme	Hm1
II	Encodes a signal transducing protein kinase	PTO
III	Encodes a receptor that transmit signal from protein kinase to phosphokinase for further amplification	Xa21
IV	Encodes cytoplasmic proteins that have a nucleotide binding site and transmit signal to nucleus	N',L6
V	Encodes protein in cytoplasmic membrane that reacts with elicitor product ofavr gene	Cf9

How do R Gene Confer Resistance?

- Elicitor molecule produced by an avr gene of pathogen in recognized by a specific plant receptor encoded by an R gene.
- Following recognition of the elicitor by the receptor molecule, one or more kinase enzymes may become activated that then amplify the signal by phosphorylating, and thereby energizing, other kinases and other enzymes.
- This lead to a cascade of biochemical reactions that results in the HR and thereby, localized host resistance at the point of attack of the pathogen.
- In many cases, hypersensitive response is followed by the development of various levels of SAR which is expressed in the vicinity of attack as well as in distinct parts of the plant.

Vertical resistance is most effective when

- VR is incorporated in annual crops that are easy to breed, such as small grains.
- Directed aginst pathogens that do not reproduce and spread rapidly, Ex. Fusarium or pathogen that donot mutate frequently like Puccinia
- Consists of strong genes that confer complete and long term protection to the plant which carries it.
- Host population does not consist of single genetically uniform variety grown over large acreages.
- If one or more of these, and several other, condition s are not meet, VR becomes short lived, that is, it breaks down as a result of the appearance of new pathogen races that can by overcome it.

- Varieties with VR are often attacked suddenly and rapidly by new vriulent race lead to severe epidemics: Various strategies have been developed to avoid these disadvantages:
- Multilines: mixture of individual varieties that are agronomically similar but differ in their resistance genes.
- Pyramiding: Mixture of varieties that contain different resistance genesand selecting from those that contain the mixture genes.

Horizontal Resistance

- Confer incomplete but more durable protection.
- Does not breakdown quickly and suddenly.
- Most physiological processes that act as mechanisms of defense and that are beyond the limits of the capacity of the pathogen to change.
- Universally present in wild and domesticated plants and operates against all races of pathogen including the most pathogenic ones.
- Varieties with HR remain resistant much longer than varieties with VR but the level of resistance in plants with HR is much lower than in plants with VR.

9

Local and Systemic Acquired Resistance

1. Induced resistance is at first localized around the point of plant necrosis, caused by infection by the pathogen or by the chemical is called LOCAL ACQUIRED RESISTANCE.
2. Subsequently, resistance spread systematically and develops in distal, untreated parts of the plant and it is called SYSTEMIC ACQUIRED RESISTANCE.
3. Several chemical compounds such as salicylic acid, archidonic acid and 2.6- dichloroisonicotinic acid may induce localized and systemic resistance in plants at levels not causing tissue necrosis.
4. Such chemicals may be effective in inducing resistance in plants when thry are applied through the roots, as a foliar spray or stem injection.

Characteristic of LAR

1. LAR results in near absence of lesions immediately next to the existing lesion.
2. Smaller and fewer local lesions developing farther not from the existing local lesions when inoculations made at least 2-3 days after the primary infection.
3. LAR limit the number and size of lesions per unit area.
4. Eaxample:

   LAR is induced in a 1to 2 mm zone around local lesions caused by TMV on hypersensitive tobacco varieties and probably in other host combinations.

Characteristic of SAR

1. SAR acts nonspecifically throughout the plant and reduces the severity of disease caused by all class of pathogens.
2. SAR has been studied most in cucurbits, solanaceous plants, legumes, and gramineous plants following infection with appropriate fungi, bacteria and viruses.

3. SAR cannot be induced after onset of flowering and fruiting in the host plant.
4. SAR is characterized by the coordinate induction in uninfected leaves of inoculated plants of at least nine families of genes known as SAR genes.
5. Products of several SAR genes, for examples, -1-3 glucanases, chitinases, PR-1 proteinns, have direct antimicrobial activity or are closely related to classes of antimicrobial proteins.
6. SAR does not affect spore germination and appressorium formation. However, penetration is drastically reduced in systemically induced resistant tissue.
7. SAR is also characterized by induction of peroxidase and lipoxygenase activities the lead to the production of fatty acid derivatives which exhibit strong antimicrobial activity.
8. Salicyclic acid and isonicotinic acid are true SAR activators because not only do they induce resistance to the same spectrum of pathogens and induce expression of same genes as do pathogens, but these chemicals have no antimicrobial activities.

Systemic Acquired Resistance (SAR) & Induced Systemic Resistance (ISR)

Features	ISR	SAR
Definition	Resistance activated in response to artificial application of non pathogenic strains of pathogen, PGPR, and abiotic stresses	Resistance activated in response pathogenic infection.
Pathway	Jasmonic acid dependent pathway	Salicyclic acid dependent pathway
Signalling molecule	Jasmonic acid	Salicyclic acid
Defense gene involved	CHIB, HEL, PDF1, 2	PR-1 and PR-2
Pathogen associated	Operates against biotrophic pathogens	Operate against necrotrophic and facultative parasite

10

Elementary Genetic Engineering

Genetic Engineering

- Referred as Gene technology, Transgenic technology, Transformation technology, and Recombinant DNA technology.
- Technology of developing transgenic organisms.
- It is a process of adding, modifying and removing desirable genes into the recipient cell or into the DNA strand of recipient plant by molecular techniques.
- It consists of tools and techniques that are used to study identify or modify the genes of living organisms.

Transgenes

- Genetically engineered or genetically modified genes are called transgenes.
- Such genes are utilized from another unrelated plant or from another organism.
- It may be from micro organisms (virus, bacteria and fungi), unrelated species or from animal.
- Thus, transgenic individuals contain some DNA from another organism.

Transgenic

- Genetically engineered organisms or individuals are called transgenics.
- Term transgenic was first used by Gordon and Ruddle in 1981.
- Plants containing transgenes are referred to as transgenics.
- A transgenic crop plant contains one or few genes which have been artificially inserted bypassing sexual process .
- Transgenic individuals contain a new gene inserted into them by gene technology.

Transgenic Technology

- Technique of developing transgenic individuals is referred to as transgenic technology.

- It enables plant breeders/plant pathologists to combine useful genes in one plant from a wide range of living organisms.
- It provides means for identifying and isolating genes controlling specific characteristics in one kind of organism, and transferring these genes into another quite different organism, which will then also have those characteristics.

Recombinant DNA Technology

- Recombinant DNA technology refer to a set of techniques that are used to isolate the DNA from one organism, modify it and incorporate into the genome of another individual or organism.
- Recombinant DNA is a hybrid or chimeric DNA which can combine with DNA of any organism.
- It is the same as genetic engineering.

Gene cloning

- Technique of genetic engineering by which a gene sequence with many identical copies is replicated.
- Identical gene sequence is isolated by using restriction endonuclease enzyme.
- This can be obtained by making a complementary DNA from m RNA template using reverse transcriptase.
- It is then inserted into a cloning vector i.e. a plasmid or bacteriphage.
- The hybrid is used to infect a cell (plant or bacteria) and replicated within the cell.
- Gene cloning is used for identification of molecular structure of gens.

Gene Sequencing

- Refers to the determination of the order of bases of a DNA molecule making up a gene.
- DNA is purified and then broken at a specific point using restriction endonuclease enzyme.Thus ,all the strands have one identical end.
- These strand are then broken at a random distance from this end so that there are strands ending on every base present.
- These strands are then separated , their end base identified and put in order of fragement size to determine the entire sequence.

Gene splicing

- Enzyme catalysed joining of DNA fragements is reffered to as Gene splicing.

- This occur in the cell nucleus and is catalysed by splicing nzymes such as ligase.

DNA Probes

- Small segments of DNA with known base sequences, origin and function are called DNA probes.
- DNA probe can be obtained by either through DNA template or can be produced by gene cloning technique.
- Microprocessor based sophisticated equipments called 'gene synthesizing machines' are available which can produce DNA probes.
- It is useful in determining the nucleotide sequence in living as well as non living systems.

Methods of Gene Transfer

Four methods of inserting foreign (DNA) into crop plants.

1. Plasmid method
2. Particle bombardment method
3. Direct DNA uptake
4. Microinjection

Plasmid Method

- This is agrobacterium mediated technique of foreign gene transfer.
- Also known as biological method of gene transfer.
- **Main features.**
- Used with cell culture or protoplast culture, hence regeneration from cells or protoplasts is a basic requirement.
- Cheaper method of transformation than gene gun method.
- It has some control on copy number and site of integration of transgenes.
- Main drawbacks of this method are host specificity; somaclonal variation and slow regeneration

Particle Bombardment Method

- In this method, foreign DNA is inserted into plant cells through gene gum using gold, or tungsten coated DNA particles.
- Also known as biolistic method, particle acceleration microballistic method.
- **Main features**
  - Method is simple than plasmid method.
  - Does not have host specificity.

- It is used with meristems, immature embryos, emrbyogenic callus, suspension and hypocotyls.
- High cost of gene gun prohibits use of this method.
- There is no control on copy number and site of foreign gene integration.

Direct DNA Transfer

- This method involves electroporation and is used with protoplast.
- Regeneration from protoplast is basic requirement of this method.
- This method has limited applications.

Main features

- Cheaper method of foreign gene transfer.
- Used with protoplast culture.
- Host specificity
- Regeneration is slow.

Micro-Injection Method

- In this method, plasmid DNA is delivered into host cells by microscopic needles, known as micro-injection.

Main features

- Used with protoplast culture.
- No host specificity.
- The regeneration is slow from protoplast.
- It is inexpensive method of foreign gene transfer.
- Out of these, first two methods, *viz.*, plasmid method and gene gun method used for development of transgenic crop plants. Other two methods are not in much use.

Genetic engineered crops that show enhanced resistance to viral pathogens

Gene	Stage of infection cycle	Mode of action	Virus	Transgenic plant
Coat protein	Uncoating	Competition of RNA	TMV	Tobacco
Movement protein	Transport	Interfere with transport	TMV PVY	Potato Tobacco
Replicase protein	Replication	Competition for enzyme		Tobacco, Potato
Antisense RNA	Translation	Blocks viral RNA and prevent translation	PLRV	Tomato
Satellite RNA	Assembly	Competes for capsids	CMV	Tobacco
Ribozyme	Translation	Cleaves viral RNA	CEVd	Tomato

Fungal Resistant Transgenic Plants

Gene/Protein product	Source	Target pathogen	Transgenic plant
Chitinase	**Bean**	***Rhizoctonia solani***	**Tobacco**
Chitinase	**Rice**	***Rhizoctonia solani***	**Rice**
Ribosome inactivation protein (RIP)	Barley	*Rhizoctonia solani*	Tobacco
Stilbene synthase	Grapevine	*Botrytis cinerea*	Tobacco
Chitinase +Glucanase PR-1		*Cercospora nicotianae* *Peronospora tabacina*	Tobacco
Osmotin	Tobacco	*Phytophthora infestans*	Potato
B-1.3-endoglucanase	Soybean	*Botrytis cinerea*	
Gene that induce artificial cell death			
Barnase and barstar		*Phytophthora infestans*	Potato
Glucose oxidase	*Aspergillus niger*	*Phytophthora infestans*	Potato
Cryptogein	*Phytophthora cryptogea*	*Erysiphe cichoracearum* *Botrytis cinerea*	Tobacco
Nettle agglutinin		*Botrytis cinerea*	Tobacco

Bacterial Resistant Transgenic Plants

Gene/protein	Origin	Target	Transgenic plant
Lysozyme	T_4 bacteriophage	*Erwinia carotovora*	Potato
Tabtoxin acetyle transferase	*Pseudomonas syringae pv. tabaci*	*Pseudomonas syringae*	Tobacco
Horothionin	Barley seeds	*Pseudomonas syringae pv. tabaci*	Tobacco
Bacterio-opsin	*Halobacterium halobium*	*Pseudomonas syringae pv. tabaci*	Tobacco
Glucose oxidase	*Aspergillus niger*	*Erwinia amylovora*	Potato
Pectate lyase	*Erwinia carotovora*	*Erwinia carotovora*	Potato
Albicidin detoxifying gene	*Pantoea dispersa*	*Xanthomonas albilineans*	Sugarcane

11

Defense Through RNA Silencing

RNA Silencing

1. RNA silencing is a type of gene regulation.
2. Serves as an antiviral defense in plants.
3. Based on targeting specific sequences of RNA and degrading them.
4. Occurs in a broad range of eukaryotic organisms, including plants, fungi, and animals.
5. While plants use RNA silencing to defend themselves against viruses, the viruses, in turn, encode proteins by which they attempt to suppress the silencing of their RNA.
6. RNA silencing was first observed in transgenic plants transformed with viral genes providing "pathogen derived resistance."
7. It was noticed then that sense orientation genes in the transgenic plant interfered with the expression of both the transgenes themselves and related endogenous genes of the plant.
8. Because of the concurrent suppression of both genes, RNA silencing was at first called "cosuppression."
9. RNA silencing is due to a process that occurs after transcription (posttranscriptional gene silencing) of the RNA and involves targeted mRNA degradation.
9. Clues of its existence came from the discovery that plants carrying viral transgenes were resistant to related strains of the virus that replicate in the cytoplasm, which meant that silencing occurs in the cytoplasm rather than the nucleus.
10. Nucleotide sequence specificity of the RNA depends on the sequence of 21–25 nucleotides of antisense RNA produced directly or indirectly from sense transgenes, or from dsRNA.
11. The dsRNA is a trigger or an intermediate in the cleaving into small (21–25 nucleotides), sense or antisense RNAs called small interfering (siRNAs).
12. siRNAs act as guides that direct the RNA degradation machinery [the RNA induced silencing complex (RISC)] to the target RNAs.

RNA Silencing (Protocol)

1. A plant or viral gene is inserted in the plant DNA where it is expressed and produces messenger RNA (mRNA).
2. The viral gene may also be able to do that without being inserted in the plant genome.
3. RNA viruses routinely produce double-stranded RNA (dsRNA), and RNA from some abnormal genes doubles up upon itself and forms "aberrant" dsRNA.
4. Both dsRNAs are cleaved by an enzyme called "Dicer" into small interfering RNAs about 21–25 nucleotides long.
5. The siRNA fragments split into individual ssRNAs and these combine with proteins and produce an RNA in duced silencing complex (RISC).
6. This complex captures mRNAs that complement each short RNA sequence.
7. RNAs with a nearly perfect match of their sequence with that of small RNA are sliced into useless small fragments.
8. RNAs with less perfect sequence matches cause the RISC complex to block the movement of the ribosomes on the mRNA so that the mRNA is not translated and does not produce a protein, thereby silencing that RNA.

RNA Silencing and its Suppression

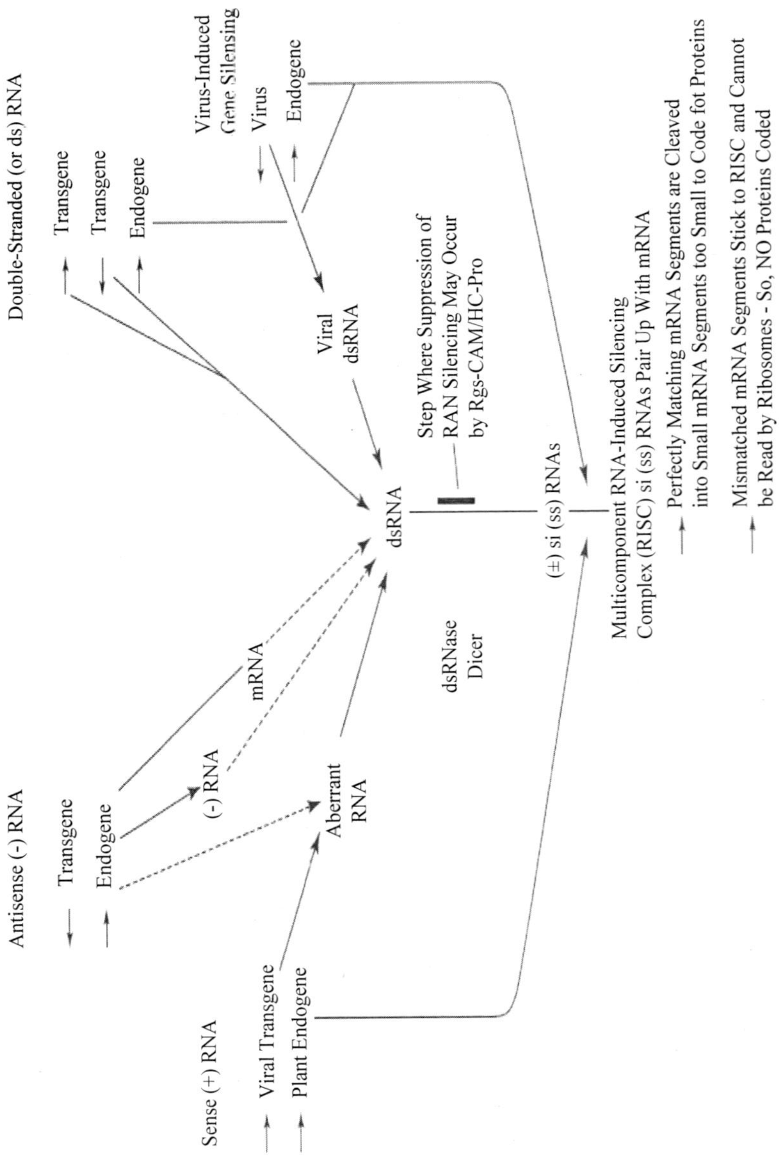

9. RNA silencing produces exceptionally strong virus resistance in transgenic plants. Such plants have neither detectable accumulation of virus in their inoculated leaves nor can this resistance be overcome with high-titer inocula.
10. Once RNA silencing of the transgene is established, all RNAs homologous to the transgene, including those from an infecting virus, are degraded.
11. Although RNA silencing is triggered locally, it can spread throughout the plant via a mobile silencing signal.
12. The movement of the silencing signal in the plant parallels that of the virus, moving at first from cell to cell and then entering the phloem and from there spreading out to parenchyma cells again.
13. The parallel movement of the virus and RNA-silencing signal may represent a race between the two, with the out-come of the race being a successful infection if the virus moves faster and becomes established first, or resistance, i.e., lack of infection, if RNA silencing becomes established first.
14. It was later shown that plant viruses could also induce RNA silencing.
15. It was further shown that virus induced gene silencing (VIGS) could be directed to either transgenes in the plant or endogenous genes of the plant.
16. As a result, plant viruses could both induce RNA silencing and could be targeted for RNA silencing by transgenes.
17. VIGS, however, is rather mild, transient, and restricted to regions around the veins.
18. RNA silencing has not yet been reported to occur in plant DNA viruses, both the ssDNA geminiviruses and the reverse transcribing dsDNA viruses.
19. All DNA viruses, however, seem to have the potential to induce gene silencing in the nucleus and in the cytoplasm, as they produce multiple copies of viral DNA genomes in the nucleus, show illegitimate integration of viral DNA into host chromosomes that mimics transgene transformation for such viruses, and generate a great deal of viral RNAs in the cytoplasm.

Suppressors of RNA Silencing

1. Soon after the discovery of RNA silencing, it was discovered that many plant viruses encode proteins that suppress RNA silencing.

2. The suppressors are structurally diverse and seem to have undergone repeated evolution steps in their attempt to keep up with developments in RNA silencing.
3. One suppressor, the helper component-proteinase of potyviruses, is so effective in suppressing viral RNA silencing that it actually increases the accumulation of several unrelated plant viruses and is, possibly, responsible for the many poty virus associated synergistic diseases of plants.
4. The same suppressor prevents both virus-induced and transgene induced RNA silencing and can even reverse an already established RNA silencing of a transgene.
5. The suppression induced by the potyvirus suppressor to a transgene induced RNA silencing can be reversed at a step at which the accumulation of siRNAs is eliminated, but it cannot eliminate the mobile silencing signal.
6. Another suppressor, *the potato virus X p25 protein, is much less* effective in suppressing RNA silencing and it apparently targets and interferes with systemic silencing.
7. RNA silencing can also be suppressed by certain host genes.
8. Some of these genes are expressed in transgenic plants, in plants following infection with certain viruses, and in transgenic plants carrying the potyvirus suppressor protein.
9. These observations suggest that the host–coded suppressor acts as a relay for the potyvirus suppressor-mediated suppression of posttranscriptional gene silencing or that the potyvirus suppressor- induced suppression of silencing perhaps takes place via activation of the host-induced suppressor protein and its unknown target protein

Unit 4: Mycology Classification of Fungi

1

Fungi; Definition, General Characteristics & Reproduction

Definition & General Characteristics

Definition

Fungus is Latin word meaning 'mushroom'

Alexopoulus and Mims 1979 defined fungus as eukaryotic , achlorophyllous organism generally produced by sexual or asexual method and whose filamentous branched somatic structure is typically covered by cell wall.

General Characteristics of Fungi

1. **Thallus:** The body of the fungus is called as *Thallus*, which is without stem, root and leaves. A single thread like filament is called as *hypha.* A group of hypha constituting the body of fungus is called as *mycelium* may be *septate* or *aseptate*. i.e coenocytic.
2. **Cell Wall:** Cell wall is well defined, typically made up of *chitin* (*Cellulose* in *oomycetes*).
3. **Nutrition:** *heterotrophic* i.e. Photosynthesis lacking and absorptive.
4. **Nuclear Status:** *multinucleate*, well defined structures i.e. nuclear membrane, nucleolus, and chromatin material etc.
5. **Sexuality:** *Asexual* and *Sexual*
6. **Life Cycle:** Simple to Complex.
7. **Sporocarps:** Microscopic or macroscopic and showing limited differentiation.
8. **Distribution:** Cosmopolitan

Somatic Structures

1. Haustoria
2. Appressoria

1. Haustoria

- A *modified mycelial branch, knob like*, makes intimate contact with plant cell, and *absorbs food.*

2. Appressoria

- These are *localized swellings* of the tip of *germ tube* that develop in response to contact with the host.
- These are special structures for attachment in the early stage of infection.

Types of Fungal Thalli

There are two types of thalli in fungi-

1. Homothallic
2. Hetrothallic

1. Homothallic Fungi

- Male and female sex organs or both the gametes are produced on the *same thallus,*
- E.g. Powdery mildew of Mung –*Sapharotheca fulginae.*

II. Hetrothallic Fungi

- If male and female sex organs or both the gametes are produced on the *different thallus*
- E.g. Rust fungi.

Fungus Tissue

There are three types of tissues found in fungi

1. Prosenchyma
2. Pseudoparenchyma
3. Plectenchyma

1. Prosenchyma

- Loosely woven tissue in which the component hyphae with elongated cells lie more or less parallel to one another is called prosenchyma.

2. Pseudoparenchyma

- Fungal tissues which are closely packed, in the form of more or less isodiametric or oval cells resembling the parenchyma cells of higher plants are called pseudoparenchyma.
- Both prosenchyma and pseudoparenchyma compose various type of vegetative and reproductive structures.

3. Plectenchyma

- Organized fungal tissues are called Plectenchyma.

Modification of Mycelium or Thallus

Rhizomorph

- Mycelium of some fungi *forms thick strands*, in such strands the hyphae lose their individuality and form complex tissues.
- Its function is transportation of water across dry areas.
- They belongs to subdivision basidiomycotina. e.g. *Arnillaria millea.*

Stroma or Stromata

- *Compact mass of the hyphae* and appears as pseudoparenchymatous tissue.
- Contain fruiting body of the fungus.
- E.g. Ergot of bajar.

Sclerotium

- Asexual, thick walled, multicelled, hard, rounded structure overseasoning structure.
- May remain dormant for long periods and germinate under favourable conditions.

Dormant Mycelium

- Mycelium which hibernates in the host tissue to tide over unfavourable conditions.
- E.g. Loose smut of wheat, Downey mildew of grape.

Gemmae

- These are the chamydospores produced in lower fungi whose walls are thinner.
- They occur either singly or in chains a becomes separated after maturing.
- Gemmae break free from the mycelium and disperse in water.

e.g. Mucor sp, Saprolegnia sp.

Reproduction

Reproduction is the formation of a progeny by either sexual or asexual means. Spore is an unit of reproductions.

Types of reproduction

1. Asexual reproduction
2. Sexual reproduction

Asexual Reproduction

This method of reproduction is characterized by production of identical individuals without the union of the sex organs.

Methods of Asexual Reproduction

1. By fragmentation
2. By arthrospores or oidia
3. By budding
4. By binary fission
5. By production of spores

Sexual Reproduction

Sexual reproduction in fungi involves the union of two compatible nuclei.

Methods of Sexual Reproduction

1. Planogametic Copulation
2. Gametangial Contact
3. Gametangial Copulation
4. Spermatization
5. Somatogamy
6. Heterokaryasis
7. Dikaryotization.

Method of Asexual Reproduction

1. By fragmentation

- Breaking off of parts of the mycelium through external forces.
- Such pieces of mycelium under favorable conditions can start a new individual.

2. By arthrospores or oidia.

- Cells of the hyphae at the distal end round off and separate in basipetal succession.
- On germination, the arthrospores give rise to new fungus colonies.

3. By Budding

- Production of a small outgrowth from a parent cell.
- Bud increases in size while still attached to the parent cell.
- It eventually breaks off and forms a new individual.
- E.g. Rust, Smut fungi & Yeast fungi.

4. Binary Fission

- Fission can occur through the simple splitting of a cell into two daughter cells by constriction.
- This is found among the bacteria generally, but some fungal yeasts may do this also.

5. Production of conidia

- Conidia produced on conidiophores in basipetal (oldest at the top and youngest at the bottom) and acropetal (oldest at the bottom and youngest at the top).

Process of Sexual Reproduction

The process of sexual reproduction consists of three distinct phases.

Plasmogamy

- It is *union of two protoplasts* brings the nuclei close together within the same cell.

Karyogamy

- Actual fusion of *two haploid nuclei* brought together as a result of plasmogamy, Karyogamy immediately follows plasmogamy.

Meiosis:

- The *nuclear fusion is followed by meiosis*. Meiosis reduces the number of chromosomes to haploid.

Methods of Sexual Reproduction

a. Planogametic copulation (Gametogamy)

- This involves the fusion of two naked gametes one or both of which are motile.
- Motile gametes are called **planogametes.**
- Depending on the size and motility of the fusing gametes, there are three types

I. **Isogamy**

- Union of gametes that are similar in shape and size.
- E.g. Synchytricum, Olpidium.

II. **Anisogamy**

- Union of gametes that are morphologically similar but differ in size.
- E.g.Genus- Allomyces.

III. **Heterogamy**

- Union between a motile male gamete with a non motile female gamete.

b. Gametangial Contact

- In this method, two gametangia of opposite sex come in contact.
- One or more gamete nuclei migrate from the male to the female through especially developed fertilization tube serves as a passage for the male nuclei.
- The zygote formed is called **Oospore**. e.g. Oomycetes, Ascomycetes.

c. Gametangial Copulation

- *Entire protoplast is transferred into the female gametangia a involes the fusion of two protoplast in a common cell.*
- Sex gametes are indistinguishable or morphological identical.
- Copualtion occurs either by complete fusion of two protoplast.
- It is common in class Trichomycetes and Zygomycetes.

d. Spermatization

- It involves the formation of small spores or seed like structures or *spematia.*
- *Spermatia acts as male gamete which is uninucleate or spore like and carried out by wind, insects to the retentive hypha* (Female gametangium).
- A pore developed at the place of contact and the contents of the spermatia passes into receptive hypha which serves as a female organ.
- E.g. Pycniospores and Receptive hypha in rusts.

e. Somatogamy

- No gametes are involved. *Vegetative hypha itself acts as a male and female gamete and bring about sexual reproduction.*
- E g. Smut Fungi.

f. Dikaryotization

- Accomplished through *migration of nuclei from one cell to another cell of vegetative hypha, often through mechanism of clamps.*
- The two nuclei remain in pair and divide as such and only fuse prior to the formation of spores.
- No special sex cells are produced.
- Clamp connections are formed during nuclear division.
- E.g. Class- Basidiomycetes.

Special type of Sexual reproduction

Parasexuality

- Some fungi do not go through a true sexual cycle as described.
- They may derive the benefits of sexual recombination through a process known as parasexuality.
- In this process, *plasmogamy, karyogamy and haploidization take place, but not a specified points in the thallus or the life cycle.*
- The parasxual cycle involves the following steps:
  1. Formation of heterokaryotic mycelium.
  2. Nuclear fusion and multiplication of the diploid nuclei.
  3. Mitotic crossing over during the division of the diploid cells.
  4. Sorting out of the diploid strains.
  5. Haplodization.

Heterokaryosis

- The phenomenon of the *existence of genetically different kinds of nuclei in the same individual is calle d heteokaryosis,* and the individuals that exhibit it are heterokaryotic.
- Heterkaryons may originate in a fungus thallus in four ways:
  1. By the introduction of genetically different nuclei into a homokaryon, a somatic cell in which all nuclei are similar;
  2. By the germination of a heterokaryotic spore, which will give rise to heterokaryotic soma.
  3. By mutation in a multinucleate, homokaryotic structure and the subsequent survival, multiplication, and spread of mutant nuclei among the wild type nuclei, and
  4. By fusion of some nuclei in a haploid homokaryon, and the subsequent survival, multiplication, and spread of the diploid nuclei among the haploid.

2

Fungi Cell Structure and Function

Key Points about fungi

1. Fungal cell walls are rigid and contain complex polysaccharides called chitin and glucans.
2. Ergosterol is the steroid molecule in the cell membranes.
3. Fungi can be unicellular, multicellular, or dimorphic.
4. Fungi in the morphological vegetative stage consist of a tangle of slender, thread-like hyphae, whereas the reproductive stage is usually more obvious.
5. Fungi like to be in a moist and slightly acidic environment; they can grow with or without light or oxygen.
6. Fungi are saprophyte heterotrophs in that they use dead or decomposing organic matter as a source of carbon.

Key Terms

- **Glucan**: Any polysaccharide that is a polymer of glucose.
- **Ergosterol**: Functional equivalent of cholesterol found in cell membranes of fungi and some protists, as well as, the steroid precursor of vitamin D2.
- **Mycelium**: Vegetative part of any fungus, consisting of a mass of branching, threadlike hyphae, often underground.
- **Hypha**: Long, branching, filamentous structure of a fungus that is the main mode of vegetative growth.
- **Septum**: Cell wall division between hyphae of a fungus.
- **Thallus**: Vegetative body of a fungus.
- **Saprophyte**: Any organism that lives on dead.

Fung: Cell Structure

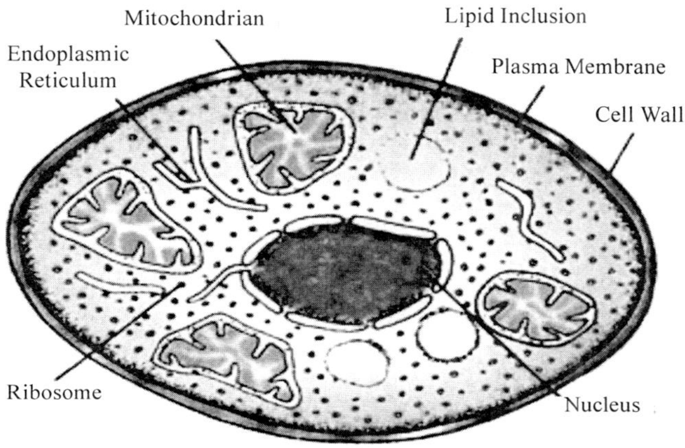

Cell Wall of the Fungus

- In the majority of fungi, the wall lacks cellulose but contains a form of chitin known as the fungus cellulose which is strictly not identical with insect chitin.
- Formula for fungus chitin is (C_{22} H_{54} N_{21}).
- In the lower fungi, the biflagellate Oomycetes are said to be distinct from all over fungi due the cellulose nature of the cell wall.
- De Bary reported true cellulose in Peronospora and Saprolegnia.
- Oomycetes may be said to have cellulose in their cell walls but it may not be the predominant material. On the other hand, glucan predominates in their walls.
- Chitin which had long been considered to be absent has recently been reported to be present even in the cell walls of some Oomycetes.
- The basic structural constituent of the cell wall in the Zygomycetes and higher fungi (Ascomycetes and Basidiomycetes) is chitin.

The Protoplast in the Fungal Cell

- The living substance of the cell within the cell wall is the protoplast.
- It lacks the chloroplasts but is differentiated into the other usual cell parts such as plasma or cell membrane, vacuolated cytoplasm, cell organelles and one or more nuclei.

Cell Membrane

- It is a delicate, extremely thin, living membrane which closely invests the protoplast.

- The cell or plasma membrane is pressed against the cell or hyphal wall except for occasional invaginations in some regions.
- Invagination is either in the form of an infolded convoluted pocket or a pouch enclosing granular or vesicular material.
- Moore and Mc Lear (1961) named it lomasome.
- Actually the plasma membrane is the surface layer of the protoplast altered to perform special functions.

Cytoplasm

- Within the plasma membrane is the colorless cytoplasm in which sap-filled vacuoles may occur.
- In young hyphae and hyphal tips, the cytoplasm appears rather uniform and homogeneous.
- Immersed in the cytoplasm are structures known as the organelles and inclusions.
- The organelles are living structures, each with a specific function.
- The inclusions are dead, have no specific function and thus are not essential to cell survival.
- Amongst the cell organelles are included the endoplasmic reticulum, mitochondria, ribosomes, Golgi apparatus and vacuoles.
- Lomasomes which are membranous structures lying between the cell wall and plasma membrane are common.
- Examples of inclusions are the stored foods (glycogen, and oil drops) pigments and secretory granules.

Endoplasmic Reticulum

- It is composed of a system of membranes or microtubular structures usually beset with small granules which are likened to the ribosomes.
- In many fungi, the endoplasmic reticulum is highly vesicular.
- Usually it is loose and more irregular than in the cells of green plants.

Mitochondria

- Cytoplasm contains small, usually spherical bodies known as the mitochondria.
- Each mitochondrion is enveloped by a double membrane.
- Inner membrane is infolded to form the cristae which are in the form of parallel flat plates or irregular tubules.
- Cristae contain the same fluid that fills the space between the two membranes.
- Mitochondria function as the power house of the cell.

- There is no fundamental difference between the mitochondria of fungi and those of green plants.
- However, Hawker (1965) holds that the cristae of fungal mitochondria are fewer, flatter and more irregular than those of the green plants.

Golgi Apparatus (Dictyosomes)

- With the exception of Oomycetes, there is less certainty of the occurrence of structures similar to those of the golgi apparatus (dictyosomes) in fungi.
- Golgi apparatus consisting of three flattened sacs surrounded by many bubble-like structures in Saccharomyces cells.

Vacuole

- Cytoplasm of young hyphae or fungal cells and hyphal tips lacks vacuoles. They appear further back or in the old cells.
- With age, they enlarge and show a tendency to coalesce and ultimately reduce the cytoplasm to thin lining layer immediately within the cell wall.

Inclusions

- Cytoplasm contains various kinds of inclusions.
- Examples of stored foods are lipid globules, granules of glycogen, oils and the carbohydrate trehalose, proteinaceous material and volutin.
- Glycogen may occur in vacuoles.
- Fungi lack chlorophyll.
- Carotenoids occur throughout the cytoplasm or concentrated in the lipid granules or distributed in the cell wall.
- Cytoplasm also secretes several kinds of ferments, enzymes and organic acids.

Nucleus

- Cytoplasm in the individual cells contains one, two or more globose or ellipsoid nuclei.
- Nuclei are small and usually range from 1-2 or 3μ in diameter.
- Structurally the nucleus consists of:
  - A central, dense body with a clear area around it.
  - Chromatin strands
  - Whole structure surrounded by a definite nuclear, membrane.
- Central body takes heavy iron haematoxylin stain and is usually Feulgen-negative.
- In electron micrographs, it appears as an amorphous or granular mass called nucleolus.
- It contains RNA.

3

Life Cycles of Fungi

Fungi

- Almost all fungi have a spore stage with a simple, haploid nucleus.
- Spore germinates into a hypha, which also contains haploid nuclei.
- Hypha may either produce simple, haploid spores again (imperfect fungi) or it may fuse with another hypha to produce a fertilized hypha in which the nuclei unite to form one diploid nucleus, called a zygote (2N).

Life Cycle of Oomycetes Fungi

- Asexual cycle (haploid spore → haploid mycelium → haploid spore) can be repeated many times during each growth season.
- Sexual cycle usually occurs only once a year.
- Zygote divides to produce diploid mycelium and spores.
- Mycelium produces gametangia in which meiosis occurs, then fertilization, and production of the zygote.

Ascomycetes Fungi

- Two nuclei of the fertilized hypha do not unite but remain separate within the cell in pairs (dikaryotic or N + N) and divide simultaneously to produce more hyphal cells with pairs of nuclei in a brief phase of most Ascomycetes.
- In Ascomycetes, dikaryotic hyphae are found only inside the fruiting body, in which they become the ascogenous hyphae.
- In these, the two nuclei of one cell of each hypha unite into a zygote (2N), which divides meiotically to produce ascospores that contain haploid nuclei.

Basidiomycetes Fungi

- Haploid spores produce haploid hyphae.
- On fertilization, dikaryotic mycelium (N + N) is produced, which develops into the main body of the fungus.
- Such dikaryotic hyphae may produce, asexually, dikaryotic spores that will grow again into a dikaryotic mycelium.

- Finally, however, the paired nuclei of the cells unite and form diploid nuclei.
- These may replicate mitotically or act as zygotes.
- The zygotes divide meiotically and produce basidiospores that contain haploid nuclei.

Deuteromycetes Fungi

- Called mitosporic fungi or imperfect fungi.
- Only the asexual cycle (haploid spore → haploid mycelium → haploid spore) is found.

4

Spore Fruit in Fungi

An aggregation of spores and spores bearing hyphae, sometimes naked but frequent enclosed

Importance Spore Fruit in fungi

1. Vital in tiding over unfavourable conditions, multiplication and maintenance of inoculum.
2. Utilized as taxonomic characters in determining the broad lines of various groups of fungi.

Type of spore fruit

1. Sexual
2. Asexual

Asexual Spore Fruits in Fungi

1. Sporangium

- Characteristic of sub-division Mastigomycotina and Zygomycotina.

2. Aecium

- Cup like or bell shaped structure producing **aeciosopores.** Eg.Rust fungi.

3. Pycnium

- Fask shaped structure containing Pyscniophores and spermatia. e.g. Rust fungi.

4. Acervulus

- Compact mass of also known as modified open sorus. e.g. *Collectrotrichum* and *Gleosporium.*

5. Pycnidium

- Closed, ostiolate fruiting body bear spores or conidia called **Pyncidiospores**. e.g. *Phoma* sp. and *Phomopsis* sp.

6. Sporodochium

- A spore fruit having cushion shaped Stroma, covered with conidiophores. e.g. Genus *Nectria, Fusarium.*

7. Synnemata

- The hyphae, which form conidiophores grouped together to form **coremia**. e. g. *Stysannus thyroseides*.

8. Sorus

- Heap like compact mass of sporophores and spore. e.g. Smut and rust.

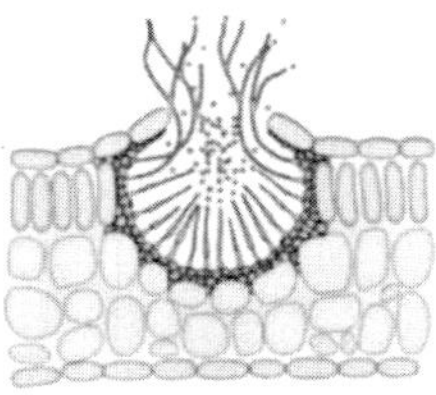

Pycnium

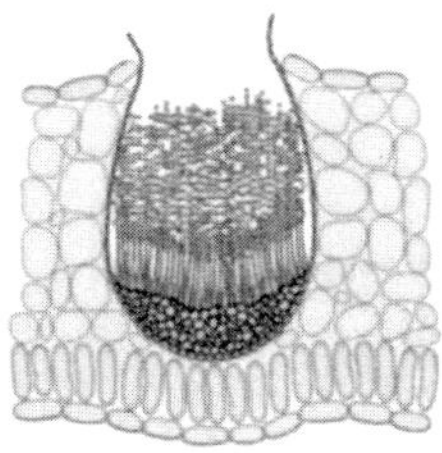

Aecium

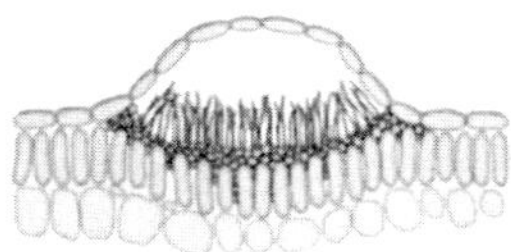

Acervulus

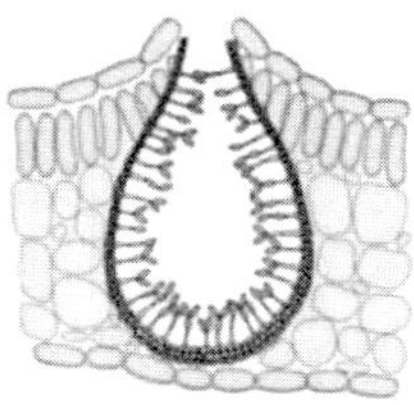

Pycnidium

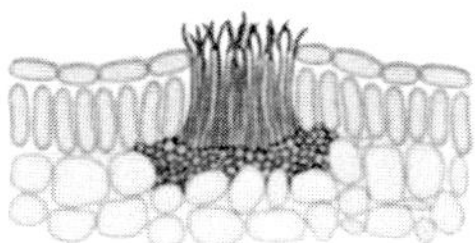

Sporodochium

Coremium

Sexual Spore Fruits in Fungi

Ascocarps

a. Spore fruit produced by the fungi of Ascomycotina.

b. Spore produced endogenously known as **Ascospores** in sac like structure called **ascus**.

Following are the different types of Ascocarps

1. Apothecium

a. An open, cuplike, or saucer-shaped fruiting body containing asci.

b. Asci are arranged in palisade layer called hymenium.

c. e.g. *Sclerotinia.*

2. Perithecium

a. Flask shaped ascocarp with ostiole.

b. Asci are arranged or lined the inner wall.

c. Sterile structures present in between the asci known as paraphyses.
d. e.g. *Claviceps* and *Glomerella.*

3. Cleistothecium

a. Closed containing asci and ascospores adapted as overwintering structure.
b. e.g. Powdery mildew fungi of order Erysiphales.

Pathogenicity

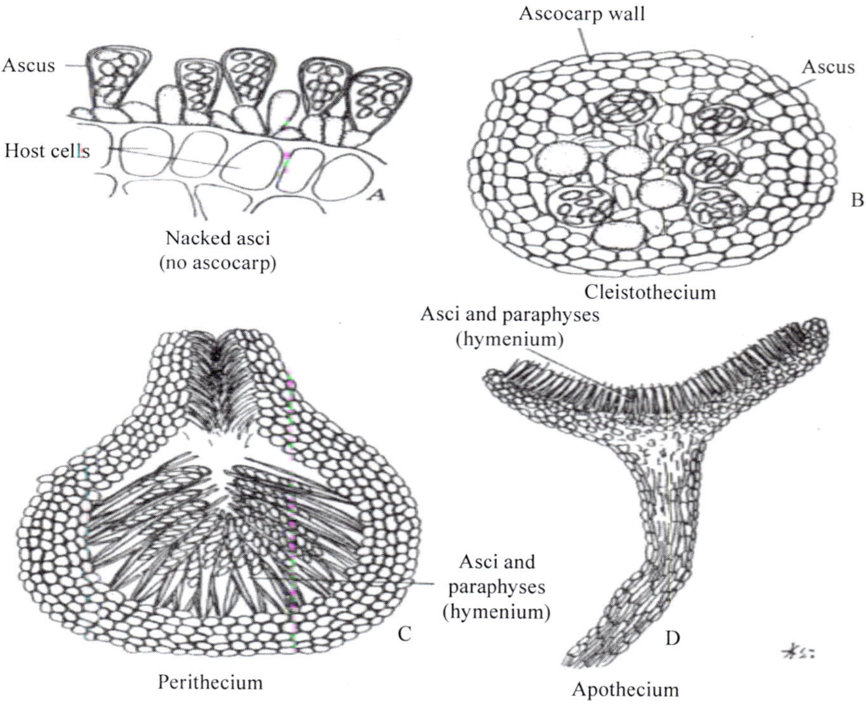

Basidiocarps

- Fructification of sub-division, Basidiomycotina.
- Consist of mushroom, bracket fungi, and puff balls.
- Sexual spores known as basidiospores which are usually 4 in number.
- Basidia are intermingled with sterile structures called paraphyses.

1. Mushrooms

- Fleshy or leathery compound fruictification with variously coloured.
- Mushroom may be edible and non-edible or poisonous.
- e.g. *Agaricus* sp.

2. Puff balls

- Round or spherical basidiocarp,
- On maturity Basidiospores are given off, in the form of puff or smoke.

3. Bracket Fungi

- Compound fructification growing on dead tree trunks called basidiocarp.

Bracket Fungus (Polyporus)

Puff Ball (Lycoperdon)

Toadstool (Amanta)

Section of Basidiocarp

5

Binomial System of Nomenclature

Terminology Used in Nomenclature

Taxonomy

- Science that deals with the identification, nomenclature and classification of organisms.

Systematics

- Scientific study of organisms with the ultimate object of characterizing and arranging them in an orderly manner.

Classification

- Ordering of organisms into groups and systems, based on their morphological, phylogenetic and physiological relationships.

Nomenclature:

- System of assigning names to the taxonomic groups or organism according to international rules.

Binomial System of Nomenclature

Rules of Nomenclature

1. Developed by Carlous Von Linnaeus which is now universally used.
2. The name of organism is composed of two words.
3. The first word designates the genus and the genus name is always capitalized.
4. The second word designates the species and its name is not capitalized.
5. Binomials when written are underlined and when printed italicized.
6. E.g. *Erwinia coli*, in which Erwinia is the genus and coli is the species.
7. The name or abbreviated name of the scientist describing the species for first time should be written after binomial. E.g. P. syringae Val Hall.
8. If the name is revised, the name of the original describer should be written in bracket followed by the name of the revising scientist. E.g. *Xanthomonas compestris* pv. *oryzae (Dye)*.

9. To avoid confusion the same binomial should not be used to name two different species.
10. The year in which organism was described should be written after the name of the author or scientist.

Classification of fungi (A case study)

The following suffixes attached to the scientific name of a taxon indicates the various ranks in the system of classification based on the one proposed by Ainsworth (1973).

Taxa	Standard endings	Description
Super kingdom	Eukaryonta	------
Kingdom	Now Fungi	A group of similar divisions
Sub-kingdom	Mycota	----
Division	mycota (suffix)	A group of similar classes
Sub-division	mycotina (suffix)	----
Class	mycetes(suffix)	A group of similar orders
Sub-class	mycetidae (suffix)	-----
Order	ales (suffix)	A group of similar families
Family	aceae	A group or collection of similar genera
Genus	---	A group or collection of similar or closely related species.
Species	---	Collection of strains having similar Characteristics
Strain	----	It is population of organism that descends from a single organism or pure culture isolate.

6

Classification of Fungi

Kingdom: Protista (Fungi), Sub-Kingdom : MYCOTA

1. Division: MYXOMYCOTA (Plasmodium present)

Sub-division: Myxomycotina

Class: Acrasiomycetes, Hydromyxomycetes, Myxomycetes, Plasmodiophoromycetes

2. Division: EUMYCOTA (Absence of Plasmodium)

1. **Sub-div: Mastigomycotina**

   Class: Chytridiomycetes, Hyphochytridiomycetes, Oomycetes
2. **Sub-div: Zygomycotina**

   Class: Zygomycetes, Trichomycetes
3. **Sub-div: Ascomycotina**

   Class: Hemiascomycetes, Plectomycetes, Pyrenomycetes, Discomycetes, Laboulbaniomycetes, Loculoasomycetes
4. **Sub-div: Basidiomycotina**

   Class: Teliomycetes, Hymenomycetes, Gasteromycetes
5. **Sub-div: Deuteromycotina**

   Class: Blastomycetes, Hyphomycetes, Coelomycetes

Subdivision: Mastigomycotina

General characters

1. Produce flagellated cells .
2. Coenocytic mycelium.
3. Cell wall contains cellulose.
4. Oospore is the sexual spores.

Classification

Four classes

1. Chytridiomycetes (Posteriorly uniflagellate zoospores)
2. Hypochytridiomycetes (Pnteriorly uniflagellate zoospores)

3. Plasmodiophoromycetes (Anteriorly biflagellate zoospores)
4. Oomycetes (Biflagellate zoospores)

Class: Chytridiomycetes

Characterized by the single posterior whiplash flagellum of their zoospores.

Three orders

Order: Chytridiales

1. True mycelium absent.
2. Rhizomycelium present in some species.
3. Fungi belonging to this order are only plant pathogenic.

Family: Olpidiaceae

Genus: *Olpidium*

Order: Blastocladiales

1. True mycelium present.
2. Sexual reproduction by planogametic copulation.

Order: Monoblepharidales

1. True mycelium present.
2. Sexual reproduction by heterogametic copulation.

Class: Plasmodiophoromycetes

Order: Plasmodiophoromycetales

Family: Plasmodiophoraceae

- All are obligate parasites
- It form a wall less, naked plasmodium as the somatic phase.

Genus: *Plasmodiophora*

Example: Club root disease of cabbage-*Plasmodiophora brassicae*

Class: Oomycetes

1. Cell wall is made of cellulose.
2. Zoospores biflagellate.
3. Sexual spore: oospores.

Order: Peronosporales

Family: Pythiaceae

- Sporangiophores similar to the vegetative hyphae or if different then of indeterminate growth.

Genus: *Pythium* and *Phytophthora*

Sl. No	*Pythium*	*Phytophthora*
2	Haustoria are absent.	Haustoria are always present.
3	Sporangiophores are indistinguishable from the somatic hyphae of the mycelium.	Sporangia developed on specialized aerial hyphae, called sporangiophores.
4	Sporangia are either terminal or intercalary.	Sporangia are always terminal.
5	Vesicle present.	Vesicle is formed only rarely.
7	Sporangia are hyphal, spherical and rarely ovoid.	Sporangia are limoniform, obpyriform or ovoid.

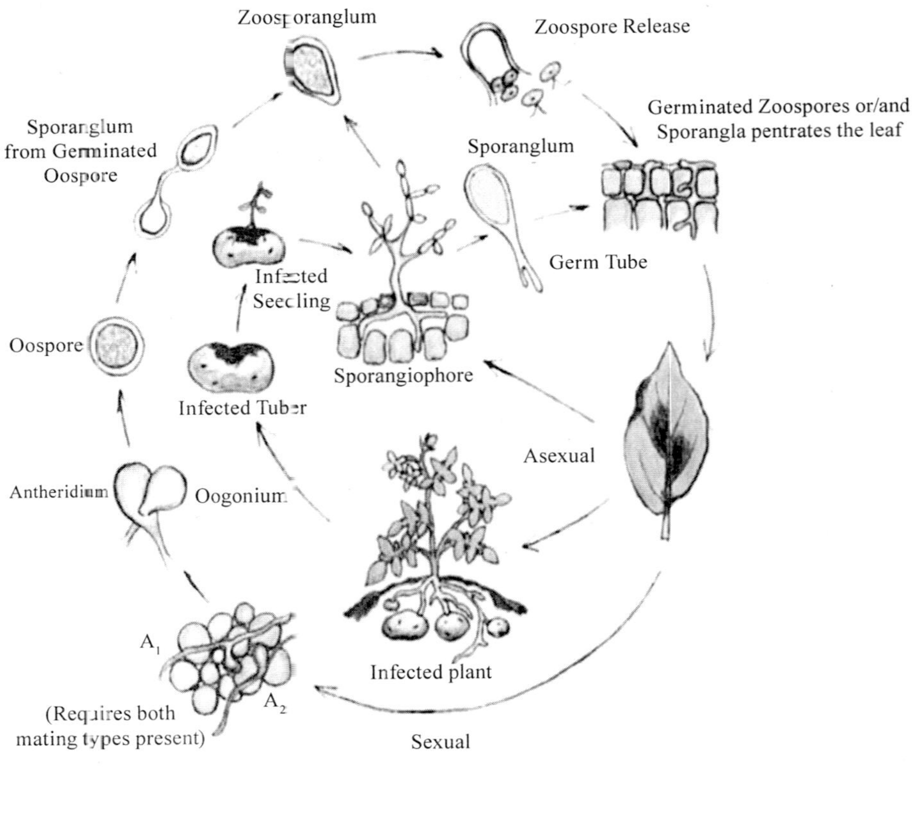

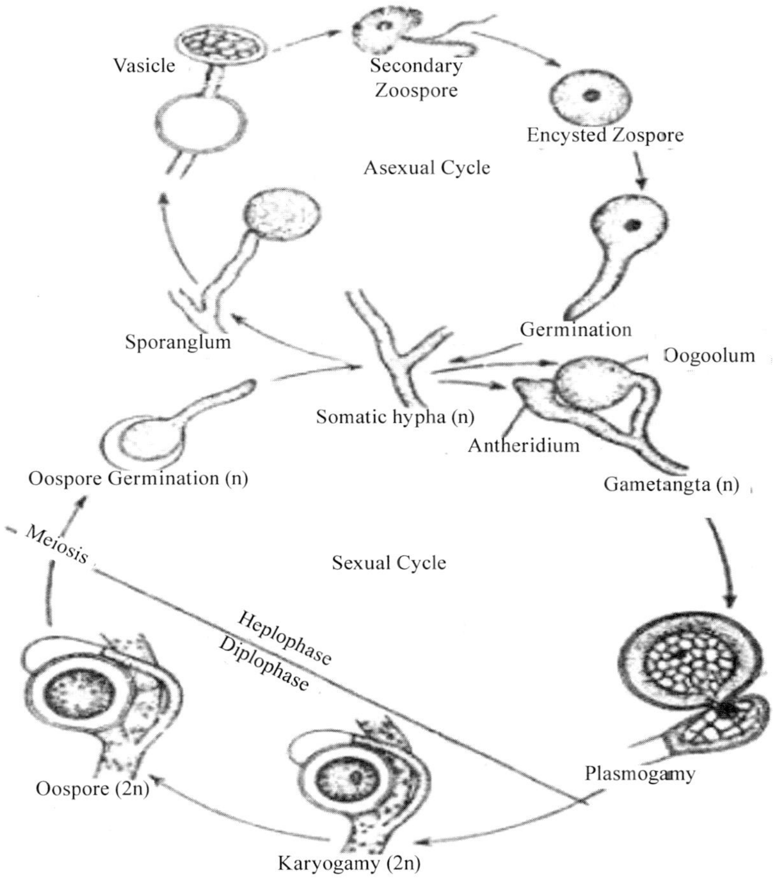

Important plant diseases caused by *Pythium* and *Phytophthora*

Fungus	**Disease**
Pythium	
P. debaryanum	Damping of tobacco and chillies
P.aphanidermatum	Soft rot of papaya, Damping off of potato
P. graminicolum	Rhizome and soft rot of turmeric
P.myriotylum	Foot rot of ginger
Phytophthora	
Phytophthora infestans	Late blight of potato
P. colocasiae	Colocasia blight
P. parasitica var. sesami	Leaf blight of sesamum
P. palmivora	Bud rot of coconut palm and toddy palm

Subdivision: Ascomycotina

General characters

1. Called 'higher fungi'.
2. Cell wall is made up of chitin.
3. Mycelium septate.
4. Asexual spores are non-motile conidia.
5. Sexual spores are ascospores usually 8 in number.
6. Ascospores are present in sac- like body, called ascus and therefore called 'sac fungi'
7. Asci are usually grouped to form fruiting body called ascocarp.
8. Ascocarps are cup or saucer shaped (apothecium), flask shaped (perithecium), or closed (cleistothecium).

Key to the classes of Ascomycotina

1. Ascocarps and ascogenous hyphae absent, thallus yeast-like -***Hemiascomycetes***
2. Ascocarps and ascogenous hyphae present, ascocarp an ascostroma - ***Loculoascomycetes***
3. Asci typically unitunicate, if bitunicate, ascocarp as apothecium: ascocarp a cleistothecium, asci evanescent and scattered - ***Plectomycetes***
4. Asci regularly arranged as basal or peripheral layer in the ascocarp ,Insect parasites - ***Laboulbeniomycetes***
5. Ascocarp perithecium, Not insect parasites, - ***Pyrenomycetes***
6. Ascocarp apothecium – ***Discomycetes***

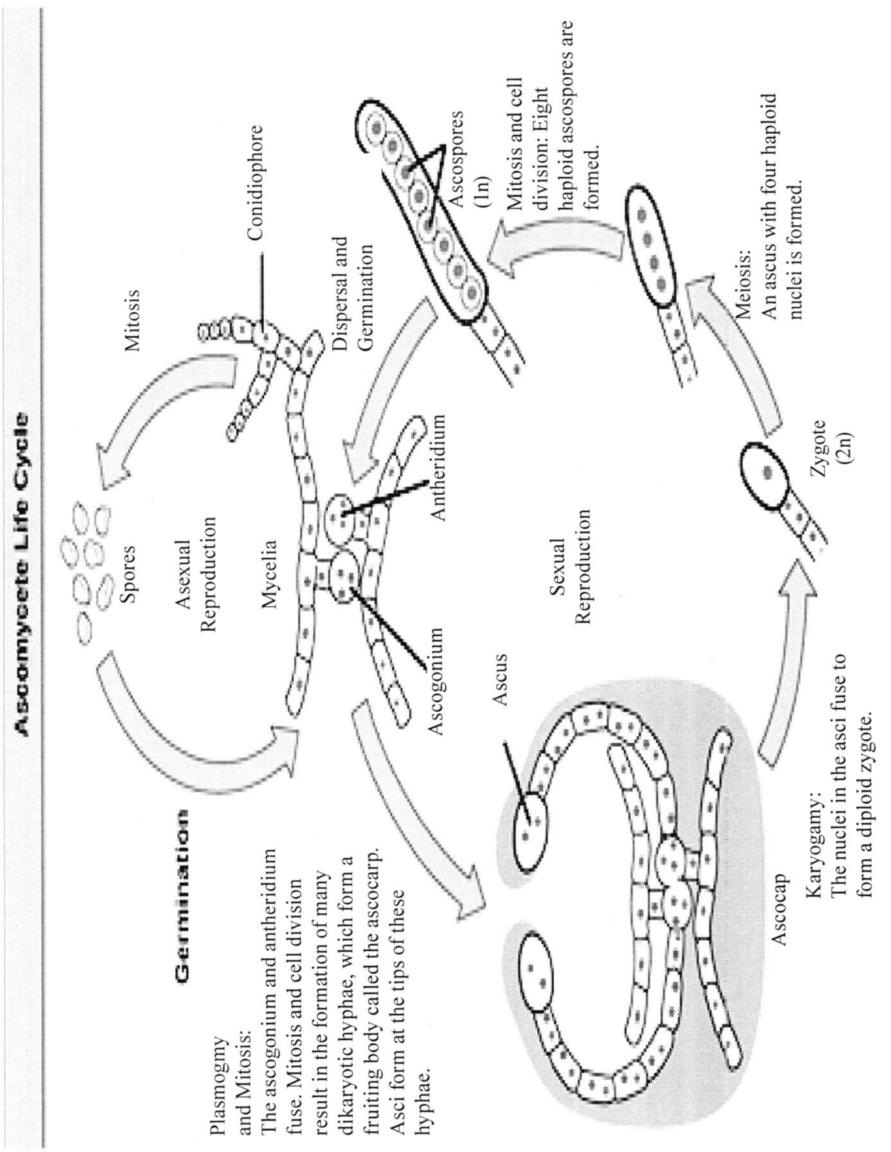
Ascomycete Life Cycle
Germination
Spores
Asexual Reproduction
Mitosis
Conidiophore
Mycelia
Antheridium
Ascogonium
Dispersal and Germination
Ascospores (1n)
Mitosis and cell division: Eight haploid ascospores are formed.
Meiosis:
An ascus with four haploid nuclei is formed.
Zygote (2n)
Sexual Reproduction
Ascus
Ascocap
Karyogamy:
The nuclei in the asci fuse to form a diploid zygote.
Plasmogmy and Mitosis:
The ascogonium and antheridium fuse. Mitosis and cell division result in the formation of many dikaryotic hyphae, which form a fruiting body called the ascocarp. Asci form at the tips of these hyphae.

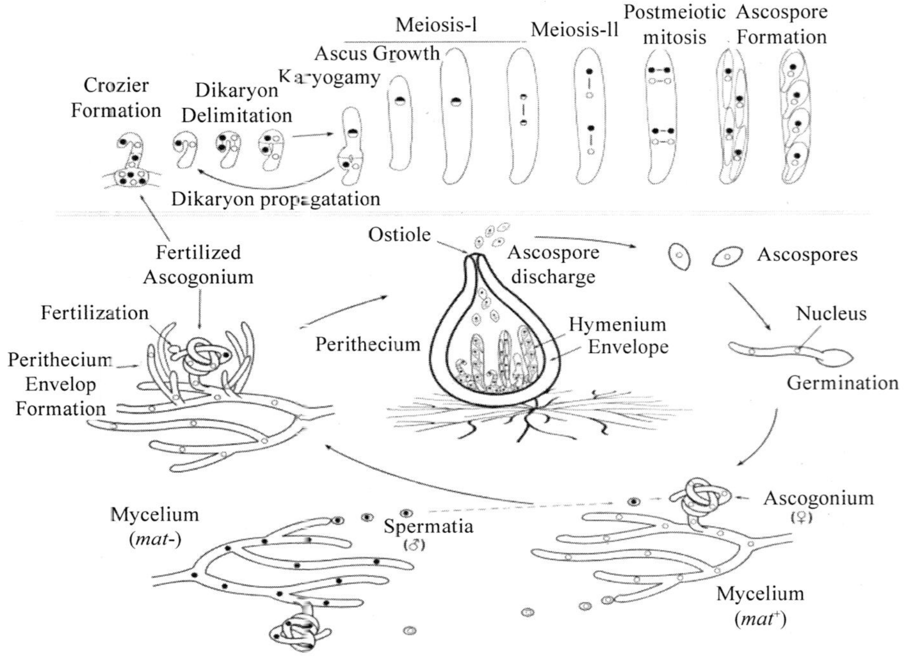
Meiosis-I
Meiosis-II
Postmeiotic mitosis
Ascospore Formation
Ascus Growth
Karyogamy
Crozier Formation
Dikaryon Delimitation
Dikaryon propagation
Ostiole
Ascospore discharge
Ascospores
Fertilized Ascogonium
Fertilization
Perithecium
Hymenium
Envelope
Nucleus
Perithecium Envelop Formation
Germination
Ascogonium
(♀)
Mycelium
(*mat*-)
Spermatia
(♂)
Mycelium
(*mat*+)

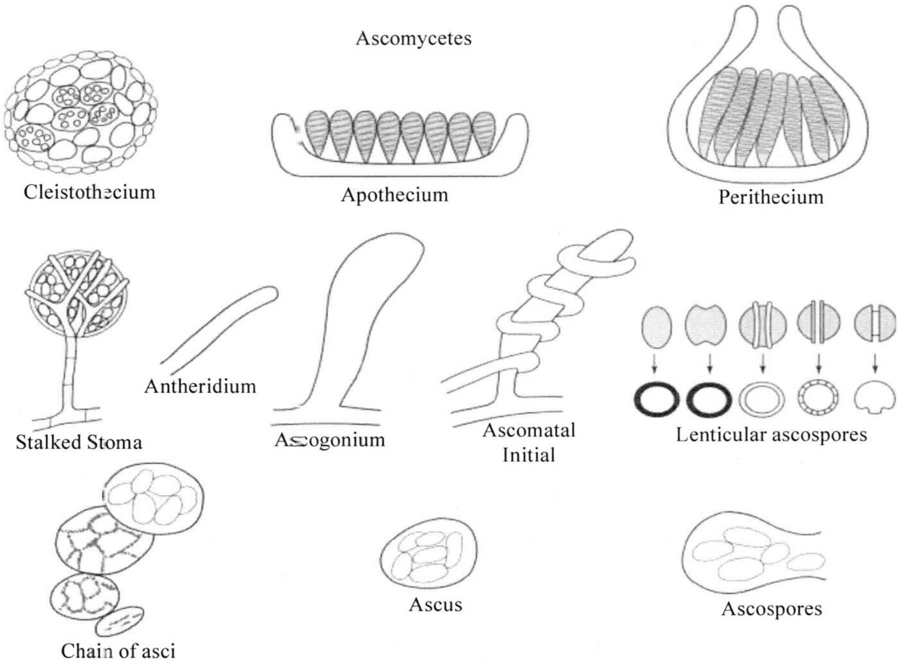
Ascomycetes
Cleistothecium
Apothecium
Perithecium
Antheridium
Stalked Stoma
Ascogonium
Ascomatal Initial
Lenticular ascospores
Ascus
Ascospores
Chain of asci

Class: Hemiascomycetes

1. Characterized by the lack of ascocarp
2. Ascogenous hyphae absent
3. Thallus yeast-like.

It is divided into three orders

I. Endomycetales

- Asci developing parthenogenetically from a single cell or directly from a zygote.

Family: Saccharomycetaceae

Genus:*Sacchromyces, Schizosaccharomyces,*

Brewer's and Baker's yeast:*Saccharomycetes cerevisiae*

II. Taphrinales

- Asci arise from binucleate ascogenous cells formed by breaking of cells from hyphae.

Family: Taphrinaceae

Genus: *Taphrina*

- Leaf curl or leaf blister of peach: *Taphrina deformans*
- Leaf spot of turmeric and ginger: *T. maculans*

III. Protomycetales

- Asci developing in ascus produced singly from thick walled chlamydospores

Family: Protomycetaceae

Genus: *Protomyces*

- Stem gall of coriander .*Protomyces macrosporus*

Class: Loculoascomycetes

Comprises the 5 orders

Order: Myriangiales

Family: Myriangiaceae

Genera: *Elsinoe, Myriangium*

Order: Dothideales

Family: Capnodiaceae

Genera: *Capnodium, Limacinia*

Family: Dothideaceae

Genera: *Mycosphaerella, Guignardia*

Order: Pleosporales

Family: Venturiaceae

Genera: *Venturia*

Fungus	Disease
Venturia inaequalis	Apple scab

Class: Pyrenomycetes

1. Characterized by unitunicate asci which arranged in a definite hymenium, usually inside a perithecium.
2. Perithecia may be globose or flask shaped.
3. In exceptional cases (Erysiphales- powdery mildews), the ascocarp may be a cleistothecium.

Order: Erysiphales

Family: Erysiphaceae

It has the following genera.

1. Ascocarps present

A. Mycelium superficial

1. **Ascocarp containing one ascus only**
   a. Perithecial appendages simple, myceloid -*Sphaerotheca*
   b. Perithecial appendages dichotomously branched -*Podosphaera*
2. **Ascocarp containing many asci**
   a. Perithecial appendages simples, myceloid -*Erysiphe*
   b. Perithecial appendages coiled at the top –*Uncinula*
   c. Perithecial appendages dichotomously branched -*Microsphaera*

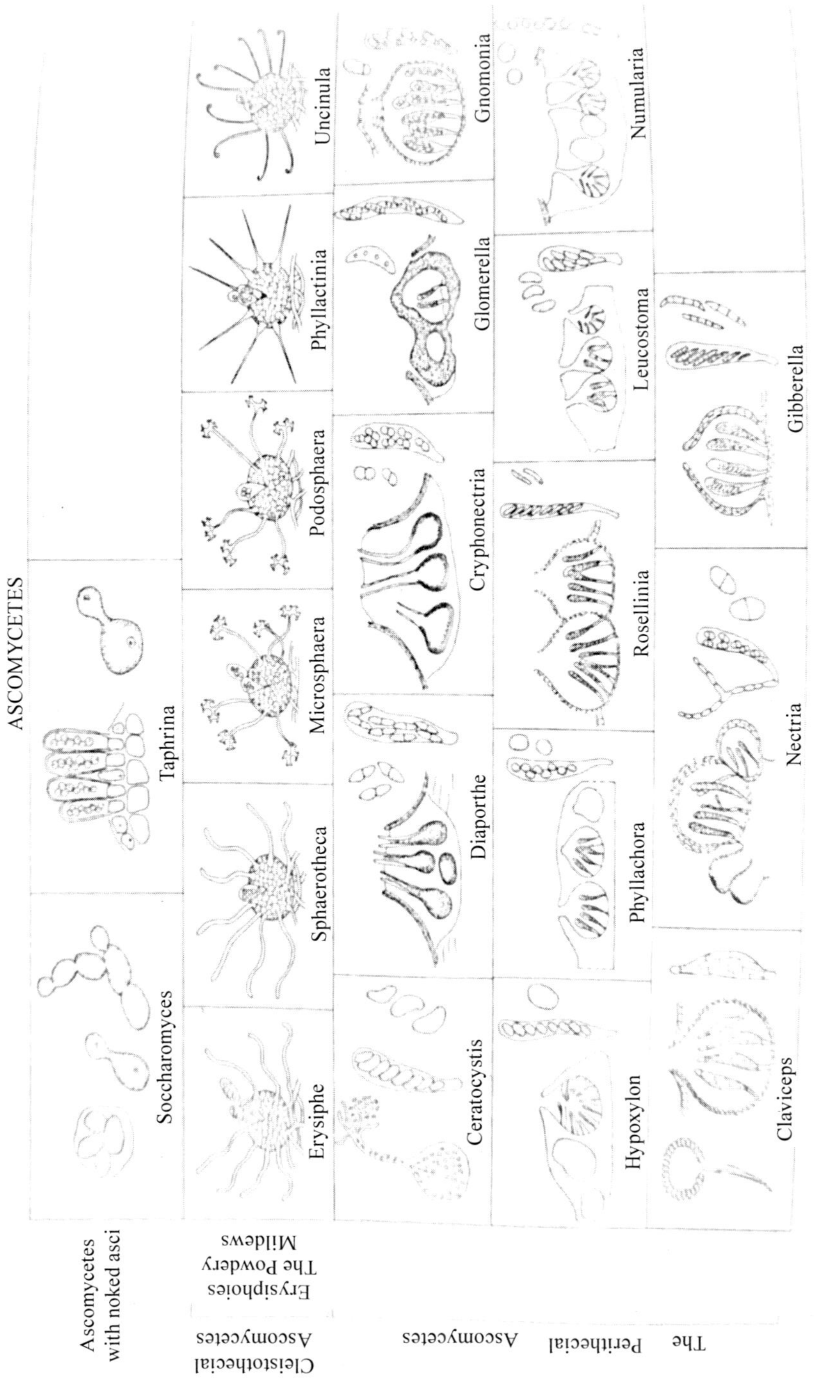
ASCOMYCETES
Ascomycetes with noked asci
Soccharomyces
Taphrina
Cleistothecial Ascomycetes
Erysiphoies The Powdery Mildews
Erysiphe
Sphaerotheca
Microsphaera
Podosphaera
Phyllactinia
Uncinula
The Perithecial Ascomycetes
Ceratocystis
Diaporthe
Cryphonectria
Glomerella
Gnomonia
Hypoxylon
Phyllachora
Rosellinia
Leucostoma
Numularia
Claviceps
Nectria
Gibberella

B. Mycelium partially endophytic

1. Perithecial appendages simple, imperfect state -*Oidiopsis*, *Leveillula*
2. Perithecial appendages coiled at the tip, imperfect state -*Oidiopsis* – *Pleochaeta*
3. Perithecial appendages with basal swellings, imperfect state – *Ovulariopsis*

II. Ascocarp absent

A. Mycelium superficial

1. Basal cell of the conidiophores swollen
2. Basal cell of the conidiophores not swollen

a. Conidia borne in chains –*Euoidium*

b. Conidia borne singly –*Pseudoidium*

B. Mycelium partly endophytic

1.Conidia ovoid, obclavate -*Oidiopsis*

2. Conidia pyriform -*Ovulariopsis*

Order: Clavicipitales

Family: *Clavicipitaceae*

Fungus	Disease
Claviceps microcephala	Ergot of pearl millet
C. oryzae	False smut of rice
Claviceps purpurea	Ergot of rye
C. sorghi	Ergot of sorghum

Powdery Mildews

1. Fungi produces closed ascocarp called cleistothecium.
2. Obligate parasites of higher plants mostly dicotyledons.

Classified as

I. One ascus in a cleistothecium

i. Myceloid appendages - e.g., *Sphaerotheca*

ii. Dichotomously branched appendages - e.g., *Podosphaera*

II. Many asci in a cleistothecium

i. Myceloid appendages - e.g., *Erysiphe ,Leveillula.*

ii. Appendage coiled at the tip (circinoid type) - e.g., *Uncinula.*

iii. Dichotomously branched appendages - e.g., *Microsphaera*

iv. Appendage with bulbous base and spear like tip - e.g., *Phyllactinia*

Fungus	Disease
Powdery mildew of grapes	*Uncinula necator*
Powdery mildew of rose	*Sphaerotheca pannosa*
Powdery mildew of wheat	*Erysiphe graminis*
Powdery mildew of shisham	*Phyllactinia corylea*
Powdery mildew of pea	*Erysiphe polygoni*
Powdery mildew of apple	*Podosphaera leucotricha*

Diseases Caused by Ascomycetes and Imperfect (Asexual) Fungi

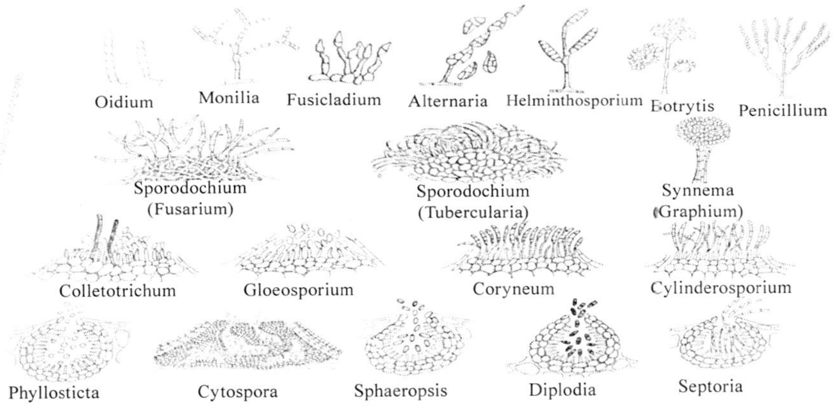

Class- Plectomycetes

- The asci are produced in ascocarp, which is mostly a cleistothecium.
- Asci are unitunicate and evanescent (i.e. dehisce at maturity so that the ascospores lie free inside the cleistothecium.

Order: Eurotiales

Family: Eurotiaceae

Genus: *Aspergillus* and *Penicillium*

Sl.No	*Aspergillus*	*Penicillium*
1	Conidiophores are unseptate and unbranched.	The conidiophores are septate and branched.
2	A conidiophores develops from cell, called foot cell.	Foot cells are absent. Con diophore develops from any vegetative cell of the mycelium.
3	Each conidiophores enlarges into a swollen vesicle at its tip.	Vesicle are not formed.
4	Metulae are not present.	Metulae are present.
5	Mature conidia are yellow, green, brown or black in colour.	Usually the mature conidia are green in colour.

Class-Discomycetes Cup fungi

1. Ascomycetes with apothecium type of fruiting bodies'
2. Discomycetes are classified into seven orders.
3. Only Pezizales and Tuberales are discussed here.

Order: Pezizales

Family: Pezizaceae (uninucleate ascospores present)

Genus: *Peziza*

Order: Tuberales

Family: Tuberaceae

Genus: *Tuber*

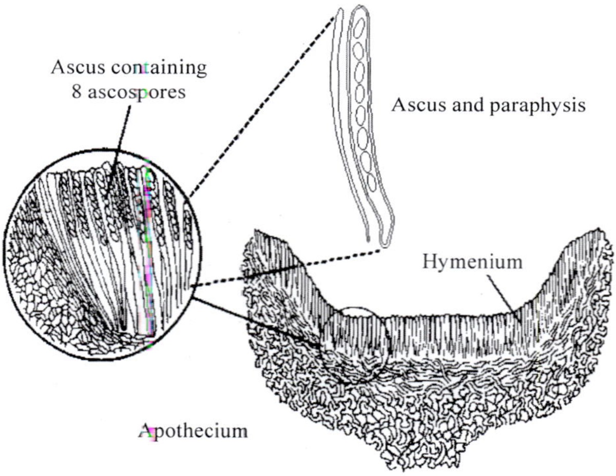

Subdivision -Basidimycotina

General characters

1. Mycelium is well developed, branched and septate.
2. Mycelium is of primary, secondary and tertiary type.
3. Dolipore septum is present except rusts and smuts.
4. Clamp connections present.
5. Plasmogamy takes place by somatogamy or spermatization.
6. Sexual spores are basidiospores (Four), exogenously produced on basidium.
7. Includes rusts, smuts, mushrooms, jelly fungi, puffballs, shelf fungi, toadstools, bird's nest fungi, bracket fungi and earth stars.

Key to the classes of Basidiomycotina

1. **Teliomycetes**
2. **Hymenomycetes**
3. **Gasteromycetes**

1. Basidiocarp lacking and replaced by teliospores grouped in sori or scattered within the host tissues - **Teliomycetes**
2. Basidiocarp well-developed, Basidia typically organized as a hymenium.
   - Basidiospores are violently discharged-**Hymenomycetes**
   - Basidiospores are not violently discharged-**Gasteromycetes**

Class: Teliomycetes

1. Basidiocarps absent.
2. Includes rusts and smuts.
3. Thick walled, dikaryotic resting spores called as teliospores in rusts and chlamydospores in smuts

This class is divided into 2 orders

1. **Uredinales**
2. **Ustilaginales**

Characters	**Uredinales**	**Ustilaginales**
Teleutospores	• Present • Formed by terminal cells of secondary mycelium	• Present • Formed by intercalary as well as terminal cells of secondary mycelium
Basidiospores	• Present in definite numbers(4) • Borne on sterigmata • Violently discharged	• Not in definite numbers • Sterigmata absent • Passively discharged
Basidiocarp	Present	Present in Graphiolaceae (Graphium)
	Rust fungi	Smut fungi

Comparison of Rusts of Smuts	
Urediniomycetes	Ustomycetes
Terminal teliospores	Intercalary teliospores
4 basidiospores produced on sterigmata	Variable number basidiospores, on sterigmata
Plasmogamy-specialized cells	Plasmogamy- no specialized cells
May require 2 hosts	Never require 2 hosts
Obligate biotrophs	Facultative biotrophs
Hosts include angiosperms, gymn sperms, gerns	Hosts include angiosperms only

Rust fungi

1. Obligate parasites of angiosperms, gymnosperms & ferns.
2. Structure of teliospores form the basis of identification of rust genera.
3. Rusts have a pleomorphic life cycle, contains five distinct spore stages that differ morpholigcally and cytologically as given below:

Stage 0: Spermagonia with spermatia and receptive hyphae

Stage I : Aecia with aeciospores

Stage II: Uredia with uredospores

Stage III: Telia with teleutospores

Stage IV: Basidia with basidiospores

1. **Autoecious rusts:**

- If all the stages produced on same host of fungus.
- Complete their life cycle in one host.

Examples

Linseed rust - *Melampsora lini* Sunflower rust - *Puccinia helianthi*

Pea rust - *Uromyces fabae* Castor rust – *Melampsora ricini*.

2. Heterocious rusts

- When all the stages produced on two host of fungus.
- Requiring two hosts for the completion of their life cycle.
- Ex. Wheat stem rust – *Puccinia graminis var. tritici.*

Types

1. Microcyclic rust

- Life cycle is short.
- Only two types of spores found.
- Teleutospores and Basidiospores.
- Example: Holly-cock *rust-Puccinia malvacearum.*

2. Macrocyclic rust

- All the five spore stages found in its life.
- Teleutospore, Basidiospore, Spermatia, Pycniospore, Aeciospore and Uredospore.
- Ex. Wheat stem rust – *Puccinia graminis* var. *tritici,* Linseed rust - *Melampsora lini,* Sunflower rust - *Puccinia helianthi,* Pea rust - *Uromyces fabae*

3. Demicyclic rust

- A macrocyclic rust in which uredospores are not formed.
- Example- Cedar apple rust - *Gymnosporangium juniperi-virginianae*

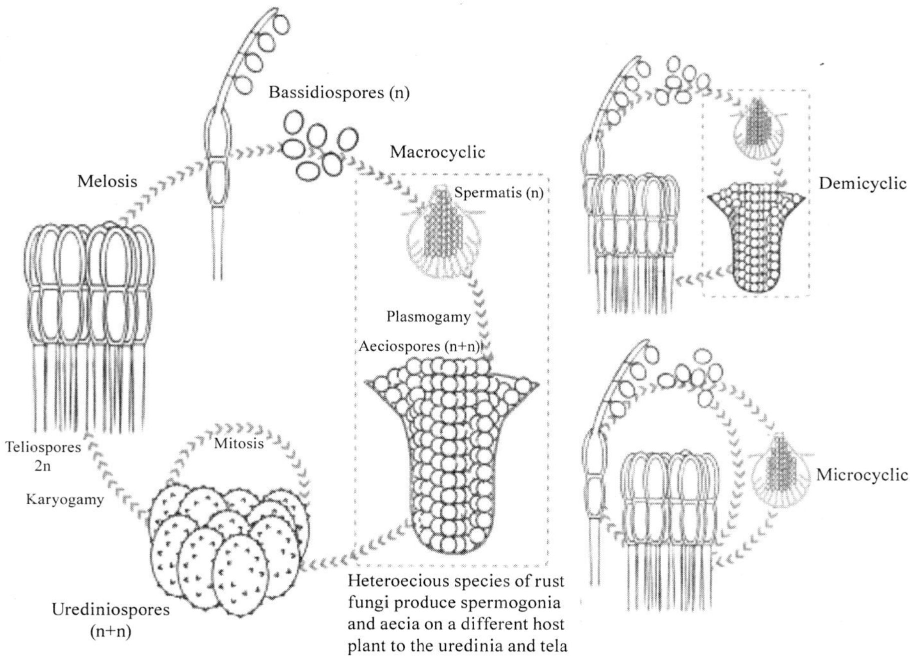

Families in Uredinales

A. Teliospores pedicellate

Teliospores germinate to form a promycelium- **Pucciniaceae**

B. Teliospores sessile

1. Teliospores germinate to produce a septate promycelium; **- Melampsoraceae**
2. Teliospores germinate without forming an external promycelium **- Coleosporiaceae**
3. Teliospores in chains **- Cronartianceae**

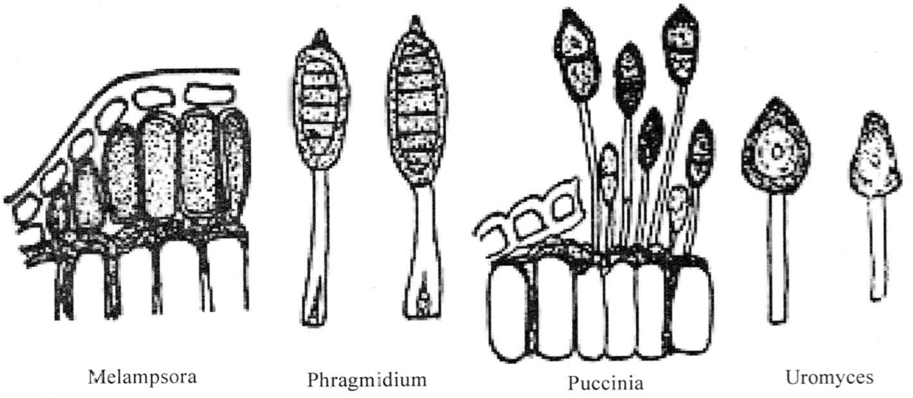

Important genera in Pucciniaceae

1. Teliospores walls colourless; uredospores reniform, basidia slender, symmetrical – ***Hemileia***
2. Teliospore walls coloured, ornamented, wall thicker above than side – ***Uromyces***
3. Teliospore cells arranged serially, with pedicel attached to the lower one only. ***Gymnosporangium***
4. Teliospores truly pedicellate, sometimes >2 celled - ***Puccinia***
5. Teliospore cells arranged as in phragmospores; wall coloured with 2 or more germ pores in each cell – ***Phragmidium***
6. Teliospore cells arranged in a radially discoid head with hygroscopic cysts – ***Ravenelia***

Different types of Teliospores

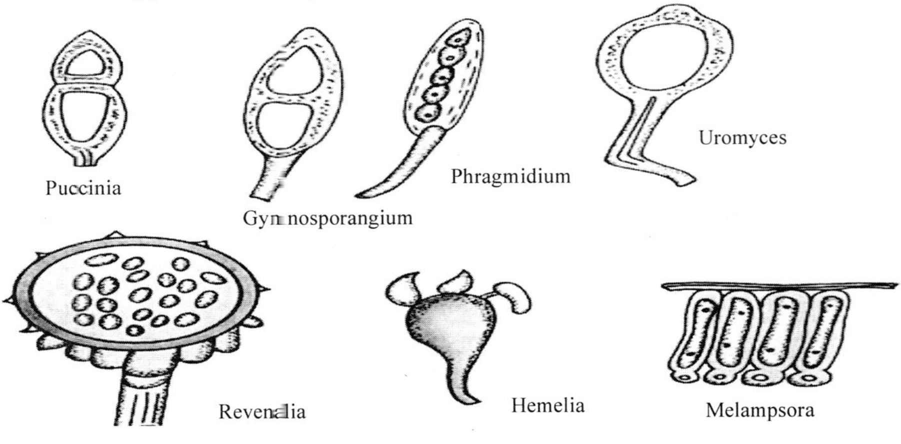

Puccinia

1. Obligate parasite and is extremely host-specific.
2. Teliospores brown with mostly 2 cells.
3. Heteroecious.

   Uredospores and **Teliospores** produced on wheat while **Pyciospores** and **Aeciospores** produced on barberry.

1. Mostly parasitize and cause rust diseases in gramineae and cyperaceae.

Important plant pathogenic species

Fungus	**Disease**
P. recondita	Brown or leaf rust of wheat
P. graminis tritici	Wheat stem or black rust
P. striiformis	Yellow or stripe rust of wheat
P. graminis hordei	Barley rust
P. graminis avenae	Oat rust
P. malvacearum	Holly hock rust
P. arachidis	Groundnut rust
P.asparagi	Asparagus rust
P. helianthi	Sunflower rust
P. hordei	Barley rust
P. arachidis	Groundnut rust

Genus: Uromyces

1. Stalked, one celled teliospores with a papillum.
2. Others characters are similar to *Puccinia.*
3. Species heteroecious or autoecious.
4. Important species causing pant diseases

Fungus	**Disease**
Uromyces ciceris-arietini	Gram rust
U. *dianthi*	Carnation rust
U. *fabae*	*Vicia* rust, Lentil rust
U. *pisi*	Pea rust

Family: Melampsoraceae

Genus: *Melampsora*

Fungus	**Disease**
Melampsora lini	Linseed rust
M. ricini	Castor rust

Order: Ustilaginales

Two families in this order.

Family: Ustilaginaceae

1. Smut fungi in which the promycelium is transversely septate with lateral and terminal sporidia.
2. Occasionally, the basidium (promycelium) develops directly into a mycelium without forming sporidia, as in *Ustilago nuda tritici,* or both conditions may be present (*Sphacelotheca sorghi*).
3. Important genera are *Ustilago Sphacelotheca*, *Tolyposporium* and *Melanopsichium*.

Ustilago

- Sori contain 1-celled teliospores, dusty at maturity and are covered by membrane of host origin.
- The sporidia germinate easily in water by infection.
- Important species causing pant diseases

Fungus	Disease
Ustilago nuda tritici	Loose smut of wheat
U. zeae	Common smut of corn (syn. *U. maydis)*
U. hordei	Covered smut of barley
U. avenae	Loose smut of oats
U. scitaminea	Whip smut of sugarcane

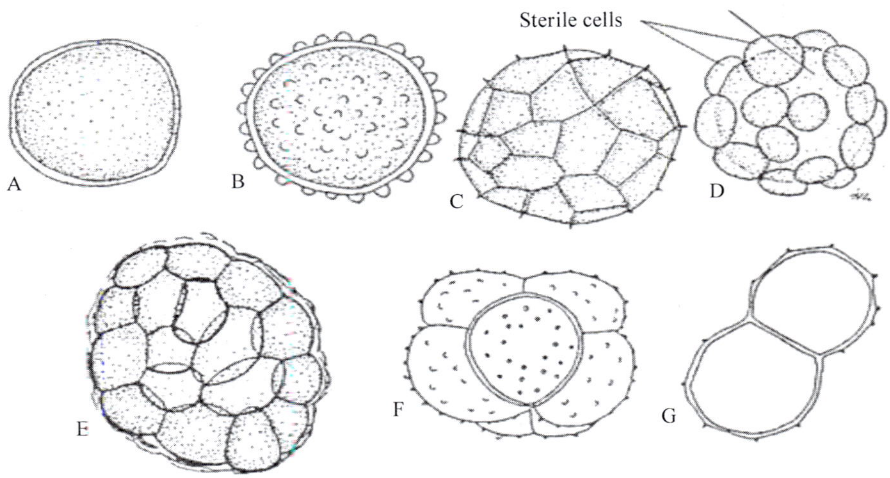

Family: Tilletiaceae

1. Smuts in which promycelium is aseptate with terminal whorl of sporidia.
2. Sporidia are longer than in Ustilaginaceae, produced in clusters at the apex of the promycelium, germinating directly into infection threads.
3. Important genera are *Tilletia, Neovossia, Urocystis, Entyloma* and *Turbicina*.

Tilletia-

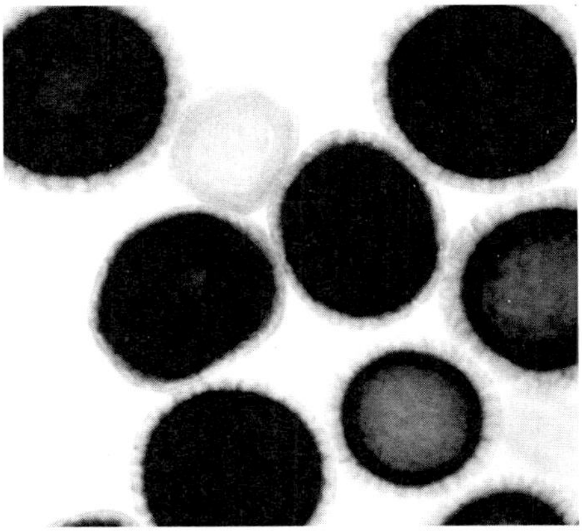

Neovossia -

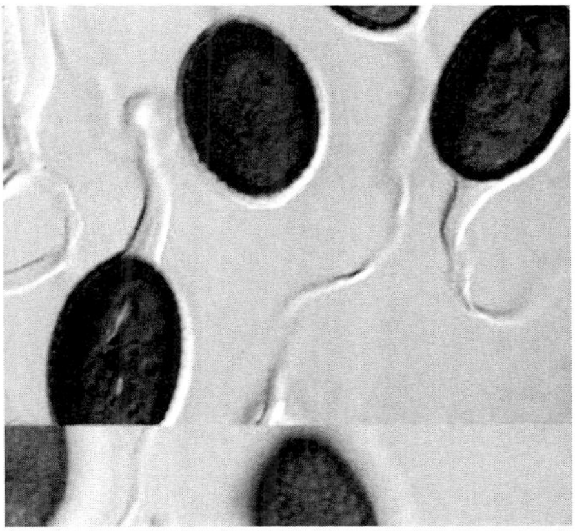

Class: Hymenomycetes

1. Characterized by usually well-developed basidiocarp or fruiting bodies.
2. Basidiocarps are typically gymnocarpic (primordium and mature sporocarp have exposed hymenium) or semiangiocarpic (partially closed till spores are matured).
3. Basidiospores are ejected forcibly i.e. these are ballistospores.
4. Well known mushrooms, jelly fungi, bracket fungi, toadstools, fairy clubs, tooth fungi, pore fungi, coral fungi

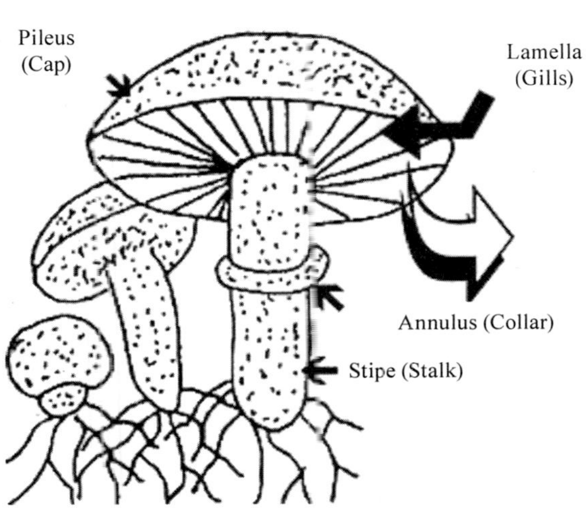

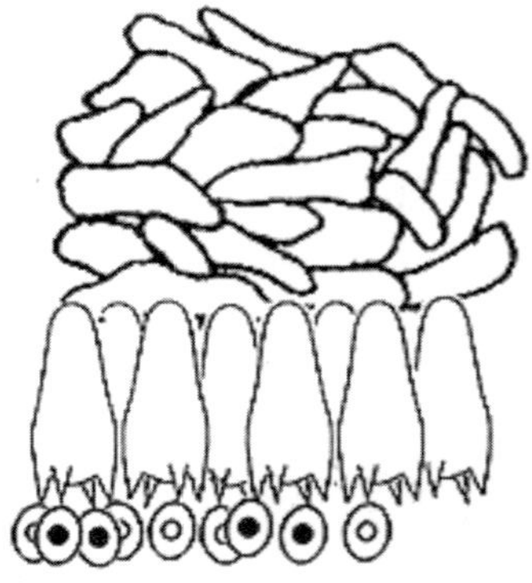

Class: Hymenomycetes

Order: Exobasidales

1. Characterized by the 4-spored basidia, which form a layer (hymenium) on the leaf surface.
2. Lack the well-define basidiocarps.
3. Consisting of the gall-forming plant parasites.

Family: Exobasidiaceae

Genus; *Exobasidium*

E. vexans - blister blight of tea.

Order: Tulasnellales

Family: Ceratobasidiaceae

Genus: Ceratobasidium and Thanatephorus

Order: Aphyllophorales/ Polyporales

Family: Corticiaceae

Genera: *Chondrostereum, Peniophora, Athelia, Corticium*

Family: Ganodermataceae

Genus: *Ganoderma*

Families in Uredinales

Order: Agaricales

1. Commonly called 'gill fungi', which include mushrooms, (edible), toadstools (poisonous) and boletes.
2. Mycelium of typically basidiomycetous with primary, secondary and tertiary mycelia.
3. Chracteristic basidiocarp or fruit body is fleshy, generally having a stalk i.e. stipulate, and has a pileus bearing hymenium-covering lamellae on the underside.

Family: Tricholomataceae

Genus: *Pleurotus, Armillariella, Marasmius,*

P. sajor -caju - Oyster mushroom

P. ostreatus - Oyster mushroom.

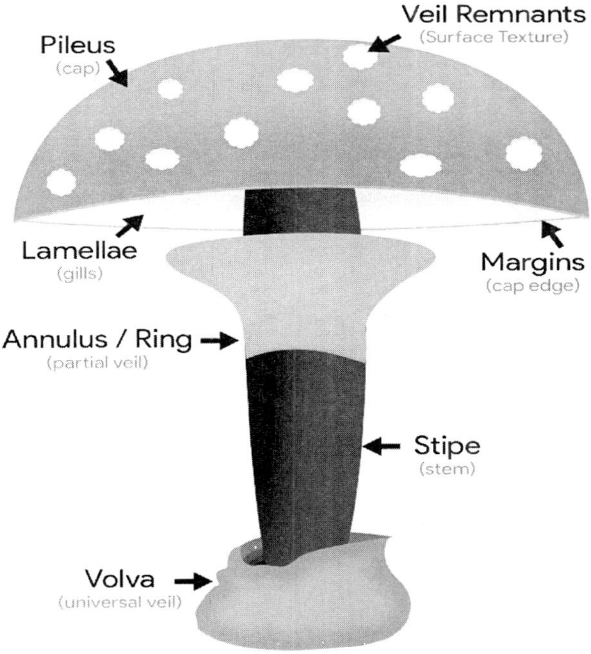

Family: *Amanitaceae*

Genus: *Amanita, Limacella* and *Termitomyces.*

Amanita

Some are more attractive and used in decoration.

Some are poisonous and produce toxins called phallotoxin and amatoxins.

A. phalloides - Called 'Depth cap fungus'and it is poisonous.

A. virosa - called as 'Destroying angel'; or death angel and it is also poisonous.

Family: Pluteaceae

Genus: *Volvariella, Pluteus, Chamaeota.*

Volvariella

V. volvacea and *V. diplasia*- commonly called the straw, or paddy straw or Chinese mushrooms are edible.

Family: Agaricaceae

Genus: *Agaricus, Cystoagaricus* and *Melanophyllium*

Agaricus

1. White button mushroom.
2. Optimum temperature for growth- 15 to 25° C.
3. Substrate: Well decomposed wheat / paddy straw compost .
4. Spawn: obtained from pure cultures of the fungus. This is called "Hyphae" permeate the compost.
5. After 2-3 weeks the compost beds are covered with a thin layer, (2.5-3cm) of soil or peat-vermiculite mixture. This is called "casing".
6. Without casing the fruiting bodies are not formed.
7. Fuiting bodies start making appearance after a month.
8. Higher temperature of the bed (15-21^{0}C) favour mycelia growth while a lower temperature (13-15^{0}C) favours fruiting body formation.
9. Temperature is kept below 15^{0}C after the initiation of fruiting.
10. 300-350 kg of mushroom can be harvested from one ton of compost in a period of 85-100 days.

Important commercial cultivated species

1. *Agaricus bisporus* -Temperate mushroom or white button mushroom,
2. *A. bitorquis*- Hot weather mushrooms

Pleurotus

1. Called Dhingri in India.
2. Called oyster mushroom as it resembles shell of an oyster.
3. Stipe is eccentric.
4. Tropical mushroom coming up well between 25-30°C.
5. Colour may be white or grey or pink depending upon the species.
6. It is grown on paddy straw (substrate) in polybags in a period of 30-45 days.
7. It yield 1.0 to 1.4 kg per kg of paddy straw.
8. Commonly cultivated species in India are
9. *Pleurotus sajor-caju, P.citrinopileatus,*
10. *P.ostreatus, P.eryingii*

Subdivision: Deuteromycotina

General characters

1. 'Perfect stage'(zygote, ascus, basidium) is either lacking or has not been discovered so far.
2. Because of the apparent absence of sexual phase, called ''Fungi imperfecti'.
3. Possess branched, septate and multinucleate mycelium except the unicellular yeast like members of Blastomycetes.
4. Reproduce only be asexual methods.
5. Conidia are produced either directly on the conidiophores or in some special types of fruiting bodies such as synnemata, acervuli, sporodochia or pycnidia.
6. Parasexuality is shown by some Deuteromycotina.

Classification Deuteromycotina

Three classes:

I. Blastomycetes

- True mycelium absent or not well-developed.
- Soma is made up of yeast (budding) cells with or without pseudomycelium.

II: Coelomycetes

- Thallus is eucarpic mycelia and septate.
- Conidia are produced either in acervuli or pycnidia.

III. Hyphomycetes

- Mycelium well developed and septate.
- Majority of the genera reproduce by conidia (Moniliales), but some members reproduce only by fragmentation e.g. *Rhizoctonia* and *Sclerotium*.
- Neither pycnidia nor acervuli are produced by any member.

Class: Hyphomycetes

Order: Moniliales

The order is divided into four families, Moniliaceae, Dematiaceae and Stilbellaceae

Family 1: Moniliaceae

- Production of free conidiophores or conidiogenous cells from the somatic hyphae and all the structures i.e. hyphae.
- Conidiophores and conidia are hyaline or light coloured.

Important plant pathogenic genera is given here

I. Conidia unicellular, globose to cylindrical, conidiophore distinct

a) Conidia almost similar to apical cells of conidiophores - *Monilia.*

b) Conidia not borne in chains; conidiophores verticillate, phialospores in mucilaginous mass -*Verticillium.*

c) Conidiophore branching irregularly or dichotomously; conidia dry, borne on inflated apicalcells -*Botrytis.*

Family 2: Dematiaceae

- Production of dark-conidia and/or conidiophores.
- Conidiophores are simple and not produced in any type of fruiting body.
- Genera: *Alternaria ,Bipolaris, Cladosporium, Cercospora, Curvularia, Drechslera, Helminthosporium* and *Pyricularia.*

Family 3: Stilbellaceae

Conidia and conidiophores develop in synnemata.

Genus: *Graphium*

Order: Tuberculariales

Characteristic features of this order is the production of sporodochia.

Family: Tuberculariaceae

Genera. *Fusarium*

Classification :-

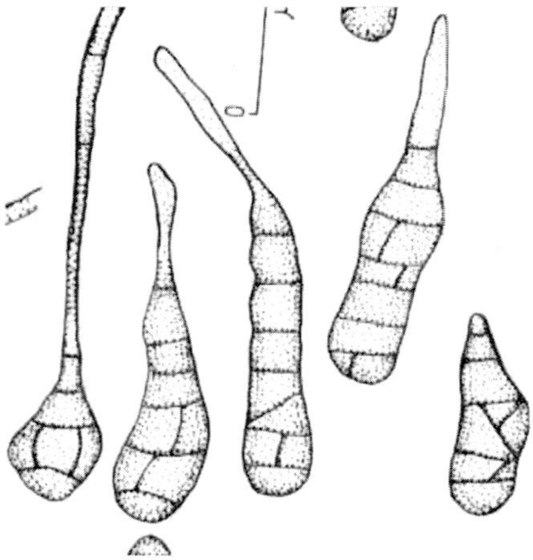

Alternaria

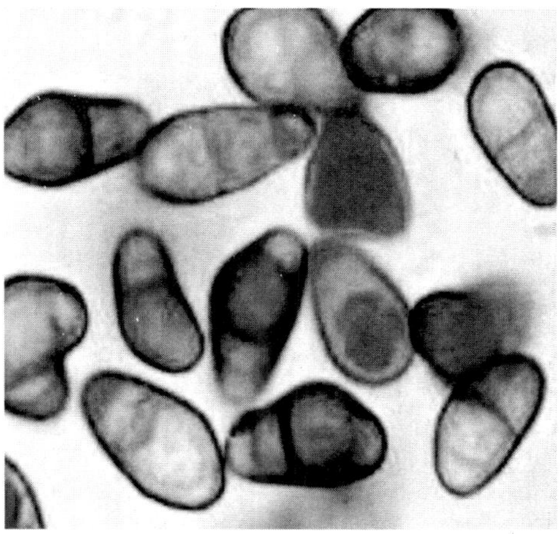

Curvularia

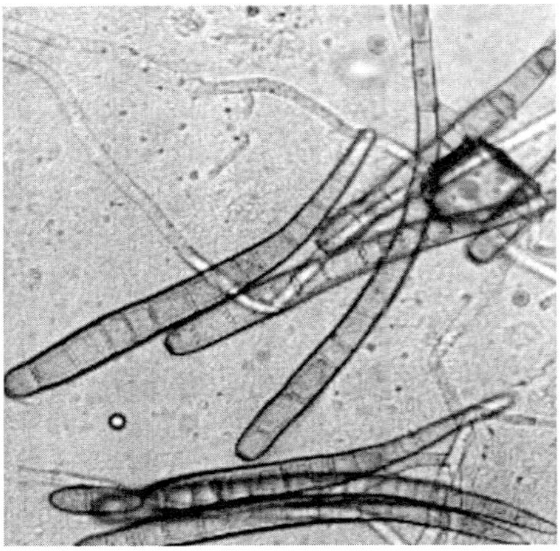

Cercospora

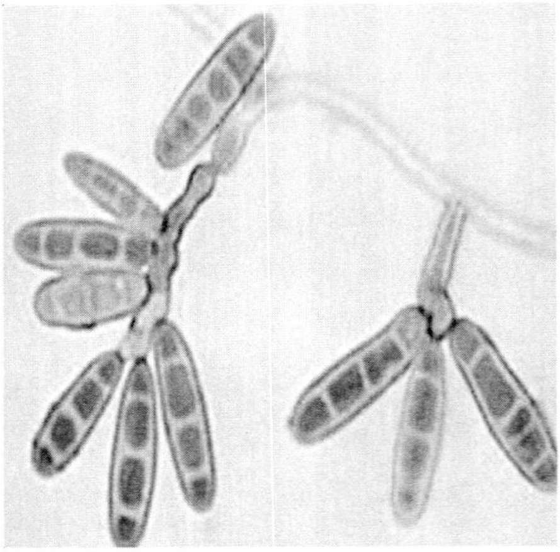

Helminothospo

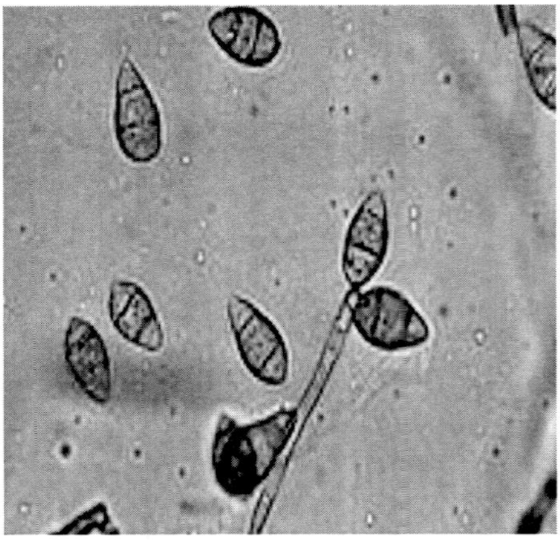

Pyricularia

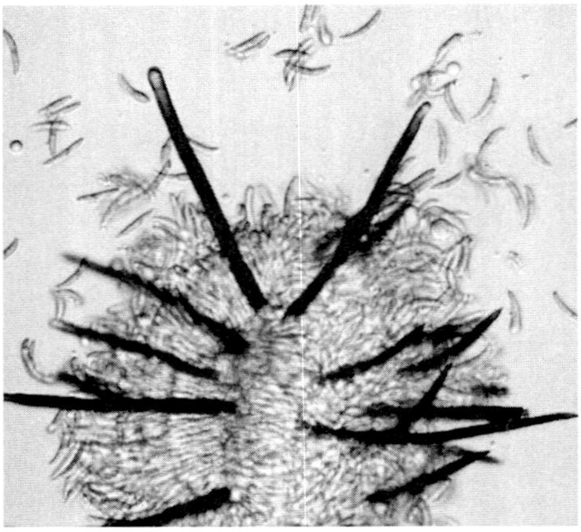

Cercospora

Class: Coelomycetes

Two orders-

1. **Melanconiales:** Conidia produced in acervuli.
2. **Sphaeropsidales:** Conidia produced in pycnidia.

Family: Melanconiaceae

1. Cause plant disease known as anthracnose.
2. Genera are *Colletotrichum, Cylindrosporium, Melanconium, Pestalotia, Pestalotiopsis, Gloeosporium*

Order: Sphaeropsidales

The conidia and conidiophores are produced in pycnidia.

Family: Sphaeropsidaceae

Genus: *Macrophomina, Ascochyta, Septoria*

Family: Excipulaceae

Genera: *Excipula, Discula, Dinemosporium, Sporonema*

Order: Agonomycetales or Mycelia sterilia

Genus: *Rhizoctonia and Sclerotium*

7

Mycorrhizae

1. Feeder roots of most flowering plants growing in nature are generally infected by SYMBIOTIC FUNGI that do not cause root disease but, instead, are BENEFICIAL to their plant hosts.
2. The infected feeder roots are transformed into unique morphological structures called MYCORRHIZAE, i.e., "fungus roots."
3. Mycorrhizae, known to be normal feeder roots for most plants, including cereals, vegetables, ornamentals, and, of course, trees and forest trees.

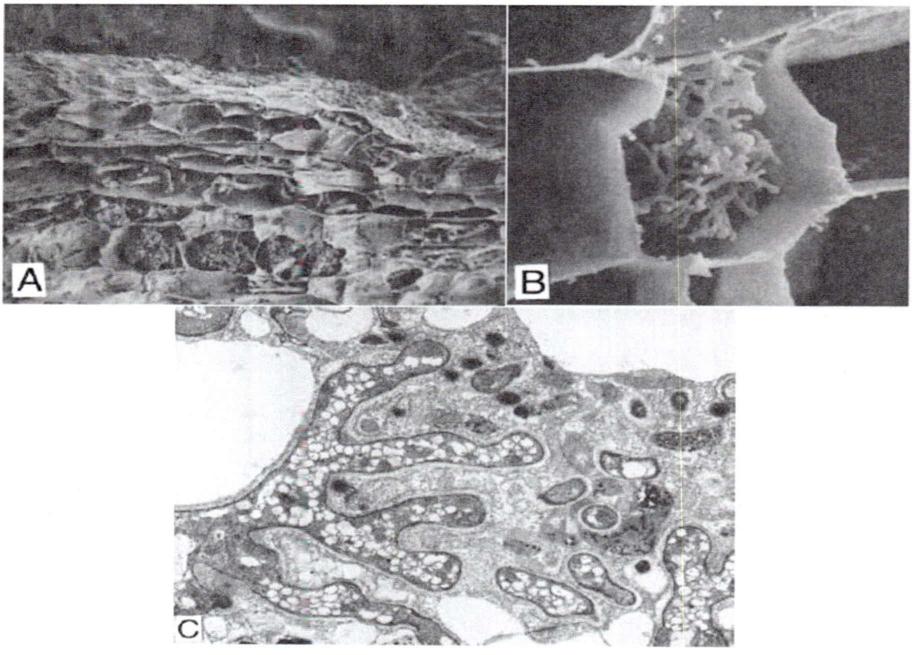

Types of Mycorrhizae

Two types of mycorrhizae, distinguished by the way the hyphae of the fungi are arranged within the cortical tissues of the root.

1. Ectomycorrhizae
2. Endomycorrhizae

Ectomycorrhizae

1. Ectomycorrhizal roots are usually swollen, appear considerably more forked than non mycorrhizal roots.
2. Ectomycorrhizae are formed primarily on forest trees mostly by mushroom and puffball-producing basidiomycetes and by some ascomycetes.
3. Spores of most ectomycorrhizal fungi are produced aboveground and are wind disseminated.
4. Hyphae of ectomycorrhizal fungi usually produce a tightly interwoven "fungus mantle" around the outside of the feeder roots.
5. Mantle varying in thickness from 1 or 2 hyphal diameters to as many as 30 to 40.
6. These fungi also enter the roots, but they only grow around the cortical cells, replacing part of the middle lamella between the cells and forming the so-called Hartig net.
7. Ectomycorrhizae appear white, brown, yellow, or black, depending on the color of the fungus growing on the root.

Endomycorrhizae

1. Most common and most important mycorrhizae.
2. Endomycorrhizae externally appear similar to non mycorrhizal roots in shape and color.
3. Internally the fungus hyphae grow into the cortical cells of the feeder root either by forming specialized feeding hyphae (haustoria), called ARBUSCULES, or by forming large, swollen, food-storing hyphal swellings, called VESICLES.
4. Most endomycorrhizae contain both vesicles and arbuscules and are, therefore, called vesicular–arbuscular mycorrhizae.
5. Endomycorrhizae are not surrounded by a dense fungal mantle but by a loose mycelial growth on the root surface from which hyphae and large pearl-covered zygospores or chlamydospores are produced underground.
6. Endomycorrhizae are produced on most cultivated plants and on some forest trees.
7. Mostly produced by ZYGOMYCETES, primarily of the genus *GLOMUS,* but also by other fungi, such as *ACAULOSPORA*.
8. *Endomycorrhizae* are also produced by some basidiomycetes.

Benefit of mycorrhiza

1. Improve plant growth by increasing the absorbing surface of the root system.
2. Alleviating water stress on plants;
   - by selectively absorbing and accumulating certain nutrients, especially phosphorus.
   - by solubilizing and making available to the plant some normally nonsoluble minerals.
   - by somehow keeping feeder roots functional longer.
   - by making feeder roots more resistant to infection by certain soil fungi, such as *Phytophthora, Pythium, and Fusarium,* and by nematodes.

Symbiosis

1. Symbiosis between the host plant and the mycorrhizal fungus is generally viewed as providing equal benefits to both partners.
2. However, it is quite probable that under certain nutritional conditions, one of the two partners may dominate and benefit more than the other.
3. Fungus is most aggressive in its invasion of root tissues when the host is growing at suboptimal nutritional levels (host defenses weak?).
4. Symbiotic relationship is terminated when nitrogen supply in the host reaches its optimum (host defenses at their best?).
5. If the nitrogen supply is again reduced to deficiency levels, the fungus partner begins to dominate and forms in abundance while plant growth is suppressed.

Mycorrhizae Vs Disease

1. Mycorrhizae do not cause disease.
2. Absence of mycorrhizae in certain fields results in plant stunting and poor growth.
3. It can be avoided if the appropriate fungi are ADDED to the plants.
4. Soil fumigation often results in the eradication of mycorrhizal fungi.
5. It causes plants to remain smaller than plants growing in non fumigated soil.
6. Systemic fungicides METALAXYL and FOSETYL-AL increase mycorrhizal infection and yield in some crop plants.

Unit 5: Plant Bacteriology

1

Identification and Classification of Bacteria

Identification of Bacteria: Plant pathogenic genera of bacteria

Agrobacterium

1. Rod shaped, 0.8 by 1.5–3 micrometers.
2. Motile by means of one to four peritrichous flagella.
3. When growing on carbohydrate-containing media, bacteria produce abundant polysaccharide slime.
4. Colonies are nonpigmented and usually smooth.
5. These bacteria are rhizosphere and soil inhabitants.

Clavibacter (Corynebacterium)

1. Cells have the shape of straight to slightly curved rods, 0.5–0.9 by 1.5–4 micrometers.
2. Sometimes they have irregularly stained segments or granules and clubshaped swellings.
3. Bacteria are generally nonmotile, but some species are motile by means of one or two polar flagella.
4. They are gram positive.

Erwinia

1. Bacteria are straight rods, 0.5–1.0 by 1.0– 3.0 micrometers.
2. Motile by means of several to many peritrichous flagella.
3. *Erwinia are the only* plant pathogenic bacteria that are facultative anaerobes.
4. Some *Erwinia do not produce pectic enzymes and* cause necrotic or wilt diseases (the "amylovora" group).
5. *Erwinia have strong pectolytic* activity and cause soft rots in plants (the "carotovora" group).

Pseudomonas

1. Pseudomonads are straight to curved rods, 0.5–1 by 1.5–4 micrometers.
2. Motile by means of one or many polar flagella.
3. Many species are common inhabitants of soil or of freshwater and marine environments.
4. Plant pathogenic *Pseudomonas species* (e.g., *P. syringae), when grown on* a medium of low iron content, produce yellow-green, diffusible, fluorescent pigments.

Ralstonia

1. Until very recently classified as Pseudomonas.
2. These resemble the latter in most respects with the important difference that its cells do not produce fluorescent pigments.

Xanthomonas

1. Cells are straight rods, 0.4–1.0 by 1.2–3 micrometers.
2. Motile by means of a polar flagellum.
3. Growth on agar media is usually yellow.
4. Most are slow growing.
5. All species are plant pathogens and are found only in association with plants or plant materials.

Streptomyces

1. Soil inhabitants.
2. Gram positive
3. Shape of slender, branched hyphae without cross walls, 0.5–2 micrometers in diameter.
4. Aerial mycelium forms chains of three to many spores at maturity.
5. On nutrient media, colonies are small (1–10 millimeters in diameter) at first with a rather smooth surface but later with a weft of aerial mycelium that may appear granular, powdery, or velvety.
6. *Streptomyces produce a* wide variety of pigments that color the mycelium and the substrate.
7. They also produce one or more antibiotics active against bacteria, fungi, algae, viruses, protozoa, or tumor tissues.

Xylella

1. Cells are mostly single, straight rods, 0.3 by 1–4 micrometers, producing long filamentous strands under some cultural conditions.
2. Colonies are small, with smooth or finely undulated margins.

3. Nutritionally fastidious, *Xylella require specialized media; their habitat* is xylem of plant tissue.
4. Gram negative, nonmotile, aflagellate, strictly aerobic, and nonpigmented.

Identification of bacteria belonging to the rod-shaped genera

1. Size, shape, structure, color, number and arrangement of flagella on the bacterial cell.
2. Chemical composition.
3. Staining.
4. Serological reactions.
5. Ability to use certain nutrients.
6. Enzymatic action.
7. Pathogen city to plants.
8. Growth on selective media.
9. Substances that they can or cannot use for food.
10. Kinds of enzymes produced on certain media.

Gram's staining reaction

1. Chemical compositions of certain substances in bacterial cells can be detected with specific staining techniques.
2. Information about the presence or absence of such substances is used for the identification of bacteria.
3. Gram's staining reaction differentiates bacteria into gram-positive and gram-negative types.
4. In this reaction, bacteria fixed on a glass slide are treated with a CRYSTAL VIOLET solution for 30 seconds, rinsed gently, treated with IODINE SOLUTION, and rinsed again with water and then alcohol.
5. Gram-positive bacteria retain the violet-iodine stain combination because it forms a complex with certain components of their cell wall and cytoplasm.
6. Gram-negative bacteria have no affinity for the stain combination, which is therefore removed by the alcohol rinse, and bacteria remain as nearly invisible as before.

Phytopathogenic bacteria: *Gram +ve Vs Gram -ve*

1. Gram positive: *Clavibacter and Curtobacterium, Arthrobacter, Bacillus, and Rhodococcus.*
2. Gram Negative*: Agrobacterium, Erwinia, Pseudomonas, Ralstonia, Xanthomonas, and Xylella.*

Symptomatology

I. External symptom

- A plant appears wilted.
- Spots on the leaves are surrounded by a halo.

II. Internal symptoms

1. Wilted plant shows discoloration of the vascular system.
2. So the wilt is caused by a pathogen and not by drought.
3. Further examination of the wilted plant can be done by placing a freshly cut wilted stem in a tube or dish of water and looking for appearance or lack of a cloudy diffusate from the stem which, if present, indicates that the wilt is cause by BACTERIA rather than a fungus or anything else.
4. If further work is needed, then one cultures the bacteria and observes the shape, size, color, and so on of its culture.
5. To make sure that the isolated bacterium is the pathogen rather than a saprophyte, a series of dilutions of the bacteria is injected into the leaves of a nonhost, such as tobacco.
6. If nonpathogenic, the leaves show no change at the points of injection. If the bacterium is pathogenic, however, it produces a hypersensitive response (dead tissues around the points of injection).

Use of Selective Nutrient Media

1. An excellent method of isolation and identification of bacteria obtained from plant tissues or soil is through the use of **selective nutrient media.**
2. **Selective** media contain nutrients that promote the growth of a particular type of bacterium while at the same time contain substances that inhibit the growth of other types of bacteria.
3. Positive identification usually requires more than one subculturing on selective media because seldom does only one bacterium grow on a selective medium.

Serological methods

1. Serological methods, especially those employing antibodies labeled with a fluorescent compound (immunofluorescent staining), are used for the quick and fairly accurate identification of bacteria and have gained popularity in recent years.
2. Use of serological methods is becoming widespread in plant pathology as the availability of species-specific and pathovar-specific antisera increases.

Quick distinction and identification of bacteria

Fatty acid profile analysis

1. In the past 10 years, fairly quick distinction and identification of bacterial genera, species, and, in some cases, lower subdivisions have been made by extraction and comparison of the FATTY ACIDS present in the bacterial cell membranes.
2. Same bacteria grown under identical conditions also produce identical membrane proteins and identical enzymes and isoenzymes. Isolation and comparison of such structural proteins or enzymes are also used to identify bacteria.

Profiles of DNA bands

1. Bacteria are detected, identified, and their genetic relatedness measured by comparison of the profiles of DNA bands obtained on a separation gel following digestion (cutting up) of the bacterial chromosomal DNA with certain restriction endonucleases.
2. Enzymes cut the DNA only at certain nucleotide sequences and release defined sets of DNA fragments called restriction fragment length polymorphisms (RFLPs).
3. RFLP profiles may be characteristic of the bacterium and, therefore, can be used to identify the bacterium.

DNA probes

1. DNA probes are used to detect and identify bacteria.
2. The probe consists of a comple-mentary segment of a part of the DNA of the bacterium that exists only or primarily in that kind of bacterium, e.g., DNA of a specific toxin gene or a virulence gene of the bacterium.
3. A radioactive element or a color producing substance is attached to the DNA probe.
4. Bacteria to be tested with the probe are treated so that their DNA is released onto a nylon membrane or filter and the DNA is then treated with the probe. If the probe finds its complementary DNA on the filter, it reacts (hybridizes) with it and stays on the filter even after washing.
5. The presence of the probe is detected by its radioactive element or the color-producing (chromogenic) compound attached to it; a positive hybridization signal, of course, indicates the identity of the bacterium tested.

Polymerase Chain Reaction (PCR) Technique

1. Availability of the polymerase chain reaction (PCR) technique, by which one or a few strands of DNA can be multiplied indefinitely to millions of copies.
2. It has made possible the detection, through a DNA probe, of the presence of one or a few bacteria in or on a seed or transplant or in a mixture of bacteria obtained from a plant or from the soil.

Classification of plant pathogenic Bacteria

Kingdom: Procaryotae

Bacteria — Have cell membrane and cell wall.

Division: Gracilicutes — Gram-negative bacteria

Class: Proteobacteria — Mostly single-celled bacteria

Family: Enterobacteriaceae

Genus

1. Erwinia, causing fire blight of pear and apple, Stewart's wilt in corn, and soft rot of fleshy vegetables
2. Pantoea, causing wilt of corn Serratia, *S. marcescens*, being a phloem-inhabiting bacterium causing yellow vine disease of cucurbits.
3. Sphingomonas, causing brown spot of yellow Spanish melon fruit.

Family: Pseudomonadaceae

Genus

1. *Acidovorax,* causing leaf spots in corn, orchids, and watermelon.
2. *Pseudomonas,* causing numerous leaf spots, blights, vascular wilts, soft rots, cankers, and galls.
3. *Ralstonia,* causing wilts of solanaceous crops.
4. *Rhizobacter,* causing the bacterial gall of carrot.
5. *Rhizomonas,* causing the corky root rot of lettuce.
6. *Xanthomonas,* causing numerous leaf spots, fruit spots, and blights of annual and perennial plants, vascular wilts, and citrus canker.
7. *Xylophilus,* causing the bacterial necrosis and canker of grapevines.

Family: Rhizobiaceae

Genus

1. *Agrobacterium,* the cause of crown gall disease.
2. *Rhizobium,* the cause of root nodules in legumes.

Family: still unnamed

Genus

1. *Xylella, xylem — inhabiting,* causing leaf scorch and dieback diseases on trees and vines.
2. *Candidatus liberobacter, phloem* inhabiting, causing citrus greening disease.

Division: Firmicutes — Gram-positive bacteria

Class: **Firmibacteria** — Mostly single-celled bacteria

Genus

1. *Bacillus,* causing rot of tubers, seeds, and seedlings, and white stripe of wheat.
2. *Clostridium,* causing rot of stored tubers and leaves and wetwood of elm and poplar.

Class: Thallobacteria — Branching bacteria

Genus

Arthrobacter, causing bacterial blight of holly.

Clavibacter, causing bacterial wilts in alfalfa, potato, and tomato.

Curtobacterium, causing wilt in beans and other plants.

Leifsonia, causing ratoon stunting of sugarcane.

Rhodococcus, causing fasciation of sweet pea.

Streptomyces, causing the common potato scab.

Mollicutes — Have only cell membrane and lack cell wall

Division: Tenericutes

Class: Mollicutes

Family: Spiroplasmataceae

Genus:

Spiroplasma, causing corn stunt, citrus stubborn disease

Family(ies): still unknown

Genus:

Phytoplasma, causing numerous yellows, proliferation, and decline diseases in trees and some annuals

2

Bacteria: Introduction, Morphology Symptomatology, Composition Survival, Dissemination & Reproduction

Introduction

1. Bacteria is prokaryotes.
2. These are generally single-celled microorganisms.
3. Genetic material (DNA) is not bound by a membrane and therefore is not organized into a nucleus.
4. Their cells consist of cytoplasm containing DNA and small (70 S) ribosomes.
5. The cytoplasm in bacteria is surrounded by a cell membrane and a cell wall.
6. Plant pathogenic bacteria have been known since 1882.
7. It cause a variety of plant disease symptoms.
8. Some types of phytopathogenic bacteria, e.g., fastidious phloem- or xylem inhabiting bacteria, which for several years were thought to be rickettsia-like organisms (RLO), were only discovered in 1972.

Characteristics of Plant Pathogenic Bacteria

Morphology

1. Most plant pathogenic bacteria are ROD SHAPED the only exception being Streptomyces, which is FILAMENTOUS.
2. Bacteria range from 0.6 to 3.5 micrometers in diameter.
3. Cell walls of bacteria of most species are enveloped by a viscous, gummy material, which, if thin and diffuse, is called a slime layer, but if thick, forming a definitive mass around the cell, is called a capsule.
4. Most plant pathogenic bacteria have delicate, threadlike flagella, considerably longer than the cells on which they are produced.
5. In some bacterial species, each bacterium has only one flagellum, whereas others have a tuft of flagella at one end of the cell (polar flagella); still

others have peritrichous flagella, i.e., distributed over the entire surface of the cell.

6. Streptomyces species, cells consist of branched threads, which usually have a spiral formation and produce conidia in chains on aerial hyphae.
7. Single bacteria appear hyaline or yellowish-white under the compound microscope.
8. When a single bacterium is allowed to grow (multiply) on the surface of or in a solid medium, its progeny soon produces a visible mass called a COLONY.
9. Colony may be circular, oval, or irregular.
10. Their edges may be smooth, wavy, or angular, and their elevation may be flat, raised, or wrinkled.
11. Colonies of most species are whitish or grayish, but some are yellow.
12. Some produce diffusible pigments into the agar that may be fluorescent with ultraviolet light.
13. Bacteria have thin, relatively tough, rigid cell walls and an inner cytoplasmic membrane.
14. Gram-negative bacteria also have an outer membrane that appears to merge with the slime layer or capsule.
15. Cell wall allows the inward passage of nutrients and the outward passage of waste matter and digestive enzymes.
16. All the material inside the cell wall constitutes the protoplast.
17. Protoplast consists of a cytoplasmic or protoplast membrane.
18. Cytoplasm, which is the complex mixture and nuclear material, which consists of a large circular chromosome composed of DNA.
19. Chromosome DNA makes up the main body of the genetic material of a bacterium.
20. Often, bacteria also have single or multiple copies of additional, smaller circular genetic material called PLASMIDS.
21. Each plasmid consists of several nonessential genes and can move or be moved between bacteria or even between bacteria and plants, as happens in the crown gall disease.

Bacterial Nutrition

- On the basis of nutritional requirements, bacteria can be divided into two categories: Autotrophic & Heterotrophic.
- Both are further divided into four major groups

Photoautotrophs	• Use light as the energy source. • Uses CO_2 as principal carbon source. • Include photosynthetic bacteria.
Photoheterotrophs	• Use light as the energy source. • Use inorganic compound as the principal carbon source. • Include purple and green bacteria.
Chemoautotrophs	• Use a chemical energy source. • Use CO_2 as the principal carbon source. • Energy is derived by oxidation of reduced inorganic compounds. • Called chemolithotrophs because of their ability to grow in strictly mineral media in the absense of light.
Chemohetero-trophs	• Utilizes a chemical energy source. • Uses an inorganic compound as the principal carbon source.

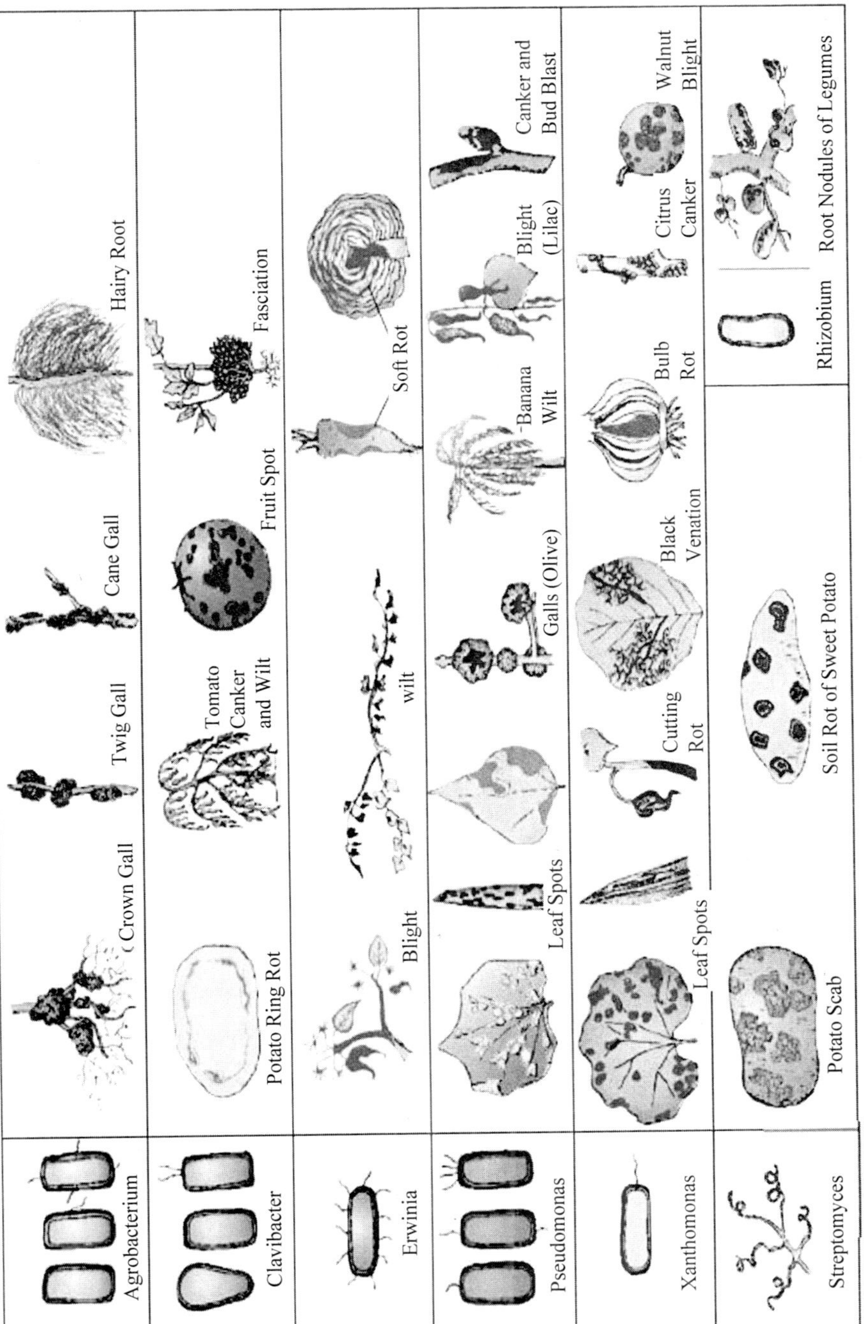

Most Important Genera of PP Bacteria and Kind of Symptoms they cause

Symptoms Caused by Bacteria

1. Plant pathogenic bacteria cause leaf spots and blights, soft rots of fruits, roots, and storage organs, wilts, overgrowths, scabs, and cankers.
2. Any given type of symptom can be caused by bacterial pathogens belonging to several genera, and each genus may contain pathogens capable of causing different types of diseases.
3. Species of Agrobacterium, however, can cause only overgrowths or proliferation of organs. However, overgrowths can also be caused by certain species of Rhodococcus and Pseudomonas.
4. Plant pathogenic species of Streptomyces cause only scabs or lesions of belowground crops.
5. Species of Rhizobium and the related genera Azorhizobium and Bradyrhizobium are gram-negative, soil-inhabiting bacteria that induce the formation of nodules on the roots of legume plants, but these bacteria are beneficial rather than pathogenic to the plant because they fix nitrogen that is used by the plants. Parts of the DNA of the three latter genera are nearly identical to parts of the DNA of Agrobacterium bacteria.

Reproduction

1. Rod-shaped phytopathogenic bacteria reproduce by the asexual process known as binary fission, or fission.
2. This occurs by the inward growth of the cytoplasmic membrane toward the center of the cell, forming a trans-verse membranous partition dividing the cytoplasm into two approximately equal parts.
3. Plasmids also duplicate and distribute themselves equally in the two cells.
4. Bacteria reproduce at an astonishingly RAPID RATE.
5. Under favorable conditions, bacteria may divide every 20 to 50 minutes, one bacterium becoming two, two becoming four, four becoming eight, and so on.
6. At this rate, one bacterium conceivably could produce one million progeny bacteria in less than a day.

Ecology and Spread

1. Almost all plant pathogenic bacteria develop mostly in the host plant as PARASITES, on the plant surface, and partly in plant debris or in the soil as SAPROPHYTES.
2. Wherever plant pathogenic, and other, bacteria exist, they often exist as BIOFILMS, that is, communities of identical or different microorganisms attached each other and/or to a solid surface.

Erwinia amyovora, which causes fire blight of pear

- Produce their populations in the plant host.
- while in the soil their numbers decline rapidly and usually do not contribute to the propagation of the disease from season to season.
- These pathogens have developed sustained plant-to plant infection cycles, often via insect vectors.

Soil Inhabitants & Soil Invaders

I. Soil inhabitants

1. Such bacteria build up their populations within the host plants, but these populations only gradually decline when they are released into the soil.
2. If susceptible hosts are grown in such soil in successive years, sufficiently high numbers of bacteria could be present to cause a net increase of bacterial populations in the soil from season to season.

Examples

1. *Agrobacterium tumefaciens,* which causes crown gall.
2. *Ralstonia solanacearum,* which causes the bacterial wilt of solanaceous crops.
3. *Streptomyces scabies,* which causes the common scab of potato.

II. Soil invaders

1. Most plant pathogenic bacteria, however, can be considered soil invaders.
2. Such bacteria enter the soil in host tissue and, because they have poor ability to compete as saprophytes, persist in the soil either as long as the host tissue resists decomposition by saprophytes or for varying durations afterward, depending on the bacterial species and on the soil temperature and moisture conditions.

Survival of Bacteria

I. In soil

1. Bacteria live mostly on plant material.
2. Less often they live freely or saprophytically, or in their natural bacterial ooze, which protects them from various adverse factors.

II. On plants

1. Bacteria often survive epiphytically, in buds, on wounds, in their exudate, or inside the various tissues or organs that they infect.
2. Bacteria may also survive in or on seeds, other plant parts, or insects found in the soil.

Dissemination of Plant Pathogenic Bacteria

1. Dissemination of plant pathogenic bacteria carried out primarily by water, insects, other animals, and humans.
2. Even bacteria possessing flagella can move only very short distances on their own power.
3. Rain, by its washing or spattering effect, carries and distributes bacteria from one plant to another and from the soil to the lower parts of plants.
4. Water also separates and carries bacteria on or in the soil to other areas where host plants may be present.
5. Insect, birds, rabbits, and other animals moving among plants may also carry bacteria

Concept of Bacterial Species

- Bacterial species are the taxonomic groups of strains.
- They are defined on the basis of common phenotypic and genotypic characteristics.

Taxospecies	• Group of strains that have a number of phenotypic characteristics in common. • They can be grouped into distinct phenotypic groups. • Ex. *E amylovora* and *R. solanacearum* are the taxospecies.
Nomenspecies	• Species with binominal names given in accordance with the rules. • Names may, therefore claimed for another taxa on the basis of some criteria. • Ex. *X. campestris* is a nomenspecies superseded by *X. comestris pv. cmpestris*.
Molecular species	• Group of strains with a high degree of nucleic acid homology. • Sequence of NA establishes and preserves the identity of species.
Genospecies	• Group of strains that can accomplish genetic exchange in some way. • Such genetic exchange occurs not only at the species level but also at genus level or even at the family level.

3

Bacterial Genetics

Rapid Development in Bacterial Genetics

Reasons

- Highly precise method of study.
- Extremely rapid growth and small size.
- Millions of bacteria are easily handled and held in a drop of liquid.
- Variety of genetic mechanisms that operate in bacteria greatly facilitated genetics analysis.
- Proved easier to related the genetic phenomena to the structure and function of the biologically important macromolecules.

Mechanisms of Genetic Transfer in Bacteria

1. Transformation
2. Conjugation
3. Transduction
4. Lysogeny

   They differ in the mode in which DNA is acquired by the recipient bacterium.

Genetic Recombination

- When the DNA enters the recipients cells by any of the three methods, it can synapse with the homologus region of the recipient genome and undergo recombination to give new genome types.
- The reciprocal product of the recombination process is broken down by DNAase enzyme.

History

1. Conjugation

- Discovered by Laderberg & Tatum (1946) in *Escherichia coli (strain K12)*
- Reported Genetic Recombination in *E. coli.*
- Awarded Nobel prize for explaining the genetic mechanism.

2. Transformation

- Discovered by F. Griffith in 1928 in *Pneumococcus pneumoniae (Diplococcus).*

3. Transduction

- Discovered by Laderberg & Zinder in 1951 started looking for recombination in *Salmonella typhimurium.*

4. Sexduction

- Discovered by Jacob & Adelberg, 1959.

5. DNA is a genetic material

- Discovered by Avery, MacLeod &Mc Carty, 1994.

Transformation –

- Strain used by Griffith in experiment with *P. pneumoniae.*

Pneumococcus strain	Morphology	Colony
Virulent	Capsulated	Smooth
Avirulent	Non –capsulated	Rough

1. Virulent strain caused death of mice on being injected while the avirulent strain was harmless.
2. Heat killed virulent strain failed to kill the mice.
3. When the living avirulent strain and the heat killed virulent strains were mixed and injected, the mice died. How?
4. Was the avirulent strain transformed into a virulent strain in assocaition with the dead virulent bacteria? YES
5. Grifith isolated bacteria from dead mice which had character of the virulent strain and proved virulent on being injected into other mice.
6. This charcter was transmitted to the progeny as a heritable character.
7. Called this phenomena as transformation and suggested that the heat killed bacteria provided the transforming principle.
8. Transforming principle was the polysaccharides of the capsule.
9. Nucleic acid as bearers of heriditory characters were not known then.

Experiment by Avery, MacLeod &Mc Carty, Proved that DNA is a genetic material

1. Identified transforming principle in 1944- DNA nd proved that DNA is a genetic material
2. They extracted DNA, polysaccharides and proteins form a killed bacterial cells in pure form and added each separately to the avirulent cultures.

3. It was only DNA that brought about the ransformation.
4. Protease and RNA degrading enymes could not prevent the transforming action whereas DNAase inactivated the transforming activity.
5. This confirmed DNA as the transforming principle.

Mechanism of Transformation

1. Fragment of DNA, present in medium, get absorbed on recipient cells and then enters the bacterial cells.
2. Inside the cell, it replaces the homologus part of bacterial genome.
3. This result in the development of new characters which pass on to the progeny cells as stable and heritable character.
4. Recipient cell should be competent having binding sites for DNA at the surface to take up the external DNA.
5. It must not have DNAase that degrade DNA.
6. Temperature and cations (Ca & Mg ions) enhance the efficiency of transformation.

Experiment by Laderberg & Tatum Suggesting Recombination in *E. coli*

1. One parental strain requires factor A & factor B for growth and was resistant to streptomycin and sensitive to phage.
2. Other parent required supply of growth factors C & D and was streptocyclins sensitive and phage resistant.
3. Individually both the strains unable to grow on minimal media.
4. When the two strains were MIXED and then spread on the same minimal medium, colonies appeared and grew in the absence of any of the growth factors A, B, C or D.
5. How did they come out?
6. Growth of colony suggests the development of recombinant bacteria which must be A+, B+, C+, D+ genetically.
7. When tested for Streptomycin and phage resistance, these markers also appeared in new recombinations.

Mechanism of Conjugation

Conjugation between F (male) & Female

1. Physical contact was involved in E coli K12 strain.
2. Strains of E. coli show sexual differences, one acting as donor of genes (male) and other as recipient of genes (Female).

3. Donor strains produce tubular F pilli which establish contact with recipient cell and serves as conjugation tube for the passage of F factor (F plasmid) and not a chromosomal gene.
4. F factor is small circular DNA molecule which is autonomous and lies free in the cytoplasm.
5. In a conjugation between F+ and F- the bacterial chromosome in not involved.
6. Only a duplicate of the F factor passes into the female cell.
7. Recipient is converted into male cell, so, the sex in *E.coli* can be said to be infectious.
8. Proved by method called interrupted mating experiment.

Conjugation between high fertility male (Hfr) & Female cells resulting in production of recombinant

- F-plasmid can live in two states–either free in cytoplasm or integrated with a bacterial chromosome.
- When inserted in the 'chromosome' the F^+ male becomes a Hfr male, 'high fertility male'.
- Hfr is a new type of male which shows enormous increase in its frequency of recombination. Super male of James Bond.
- When a Hfr male conjugates with a female F-, the genetic material is transferred.
- However, female is not converted into a male as the F^+ factor is not passed on.
- Genetic material may replace homologues portions of female genome and brings about 'GENETIC RECOMBINATION.

Transduction –

1. When a bacteriophage (phage) replicates in a bacterial cell (donor) , a few of the progeny virions incapsidate a piece of host DNA , instead of, or in addition to their own DNA.
2. These viruses when infecting a recipient host cell, introduce the DNA of the host also.

Depending on the nature of DNA introduced, there are two types of transduction

1. Generalized transduction		**2. Specialized transduction**
Any bacterial gene that is encapsulated during the lytic cycle can be transferred to the recipient and the phage genome contains only donor DNA.		Only specific genes (that lie adjacent to the inserted phage genome in the lysogenic bacteria) are transferred.

Specialized transduction

1. Caused by temperate phages which integrate in the host chromosome.
2. Bacterial genes close to prophage can be transduced (transferred).
3. Lambada phage of *E.coli* has been studied extensively for the specialized transdution.
4. During lysogenic cycle the phage genome is integrated with the bacterial chromosome.
5. When lysogenic cycle enters the lytic cycle, the phage genome is excised from the bacterial chromosome.
6. Occasionally, one or more bacterial genes that were adjacent to the viral genome go with the viral DNA.
7. After lytic cycle, the new progeny of temperate phage containing the bacterial genes enter the lysogenic cell and bring about transduction.

Sexduction

- During the dissociation of the F factor from chromosome, it may carry some of the bacterial genes such as mutant fertility factors is designated as F′ (F Prime).
- When such F′ factors as transferred, the recipients bacterial cell become heterozygous for that part of the DNA which the F′ has obtained the male bacterium.
- Phenomenon is called sexduction.
- It is employed to obtain partial diploid.

4

Growth Cycle of Viruses

Events which take place during virus growth cycle

1. Adsorption of the virion on the host cell surface.
2. Entry of the nucleic acid or the whole particle (in animal virus) into the cell.
3. Transcription and translation.
4. Genome replication.
5. Assembly of new virions, and
6. Release of virions from the cell.
7. Whole process is called virus growth cycle.

Bacteriophage

- Viruses that infect bacteria are called bacteriophages.
- Bacteriophages are shaped like tadpoles with round, oval or polyhedral heads and a hollow tail with specialized fibrils at their tip.
- Head is composed of DNA surrounded by a protein membrane which covers the tail.
- Protein coat gives protection to the bacteriophage in its extracellular environment.
- Tail fibres attach to the cell wall of the bacteria.
- Tail fibres secrete an enzyme which penetrates the cell membrane to permit entry of the virus DNA.
- DNA enters through a pore and induces the bacterial cell to manufacture more virus particles
- When bacteriophage particles multiply quickly, the bacterial cell dies.
- At the death of the cell, the bacteriophages are released & infect new bacterial cells.
- Bacteriophages survive only in bacterial cells and in the absence of them they perish.

Growth Cycle of Bacteriophages

Bacteriophages, on the basis of their growth cycles, lytic or lysogenic, can be divided into two groups:

1. Virulent phage
2. Temperate phages

I. Virulent phage

- Called a lytic cycle in which the sensitive bacterium lyses and a large number of newly-formed virus particles are liberated.

II. Temperate phage

- Called lysogenic cycle.
- Virus does not multiply, and there is no death of host cells.
- At a critical time, called induction, the phage enters the lytic cycle.

Lytic cycle

Cycle occurs in *T* series of phages which attack *E. coli*

Sequence of events of lytic cycle

Adsorption

1. Phage adsorbs on specific 'receptor sites' on the bacterial wall with the help of tail fibers and the basal plate.

Injection

1. Small fibers (tail pins) hold the phage in position and a pore is dissolved in the wall (probably by phage enzyme).
2. Then, by the contractile action of the sheath, the hollow central tube penetrates through the wall and the DNA is injected into the cell.

Protein synthesis

1. Phage-DNA 'takes over' the protein synthesis machinery of the cell and forms viral messenger RNAs and through them the proteins.
2. Some proteins are used as enzymes for the viral DNA synthesis. Proteins of the coat are also synthesized.

Genome synthesis

1. Viral DNA by replication gives rise to several copies of the viral genome for the production of progeny.

Maturation and release

1. Genomes are wrapped by the protein-coat parts to form mature phages.
2. At a certain critical time the wall of the host bacterium lyses and about 200 mature phages per cell are liberated This is LYSIS.

Lysogenic cycle

1. Occurs in X (Lambda) phages attacking *E. coli.*
2. This type of growth cycle was discovered by Andre Lwoff.
3. Phages that show lysogenic cycle are called temperate phages and the bacteria in which this occurs are called as lysogenic strains. The process is called lysogeny.
4. Cells do not lyse and no virus particles are formed.
5. A kind of 'symbiotic' association develops which is called lysogenic state.

Temperate Phage &Lysogeny

- Temperate phages differ from virulent phages.
- When phage DNA injected into the cells is integrated in to the host chromosome, & thereafter it replicates synchronously with the host genome.
- Phage genome in this state is called PROPHGE.
- Phages that have the capacity to undertake a prophage state are called TEMPERATE PHGE.
- Such relationships occurring between temperate phage and the bacteria is called LYSOGENIC BACTERIA.
- Lysogenic bacteria successfully transfer their prophage DNA to the daughter cells during cell divisions without releasing any free phages.
- Prophage is however, occasionlly realeased from the host chromosome in some host cells and initiates the lytic cycle as in virulent phages.
- This is called Spontaneous induction.

5

Structure of Bacteria

A bacterial cell consists of a

- Cell wall
- Compound membrane enclosing protoplasm
- Nucleus
- Vacuoles
- Mesosomes
- Ribosomes
- Polysaccharides & lipids
- Metachromatic granules
- Spores
- **Externally bacterial cells may show**
    - Capsules
    - Fimbriae
    - Pili
    - Flagella

Silme Layer and Capsule

- Some bacteria form an extracellular, slimy or gelatinous, polysaccharide like material as the outermost layer.
- This material may remain firmly adherent as a discrete covering layer of each cell, called a capsule.
- It may part freely from the cell as it is formed, called a free slime.
- Capsule is a part of the bacterial cell while the slime is a secretion.

Cell Wall

- Bacterial cell contains a characteristic cell wall.
- Cell wall of bacteria is composed of a peptidoglycan.

- Peptidoglycan is also called mucopeptide or glycopeptide or murein.
- Murein is composed of six different components viz. N-acetylglycosamine, N-acetylmuramic acid, L-alanine, D-alanine, D-glutamic acid and either L-lysine or meso-diaminopimelic acid.
- Rigid peptidoglycan layer is located between the cytoplasmic membrane and an outer multiple tract layer.
- The latter layer is composed of lipoprotein and lipopolysaccharide complexes.
- It is common in gram-negative bacteria.
- The external multitrack layer, the rigid peptidoglycan layer and the cytoplasmic membrane constitute the bacterial envelope.

Cytoplasmic Membrane

- Present immediately below the cell wall.
- Semipermeable.
- Controls passage of nutrients of waste products into and out of the cell.
- Membrane contains about 75 per cent protein, 20 to 30 per cent lipid and about 2 per cent carbohydrate.

Mesosomes & Cytoplasm

Mesosomes

- Formed by an invagination of the cytoplasmic membrane.
- Serve for compartmentalization and integration of biochemical systems.

Cytoplasm

- Cell material contained within the cytoplasmic membrane.
- It can be divided into the
  - Cytoplasmic area rich in ribonucleic acid (RNA).
  - Chromatinic area (nuclear area) rich in deoxyribonucleic acid (DNA).
  - Fluid portion with dissolved nutrients.

Nucleus & Ribosomes

Nucleus

- Bacteria do not have characteristic nucleus.
- They contain bodies within the cytoplasm.
- Bodies are regarded as a nuclear structure.
- DNA is confined to this area.
- Since, it is not a discrete nucleus, it is called 'chromatin body.

Ribosomes

- Globular structures found in the cytoplasm.
- They are composed of about one-third protein and two-thirds RNA.
- Designated 30s, 50s, 70s, etc. depending upon their size.
- Act in clusters and they are called polyribosomes or polysomes.
- Polysomes may contain 3 to 70 ribosomes depending upon the bacteria.
- Ribosomes are the sites of protein synthesis.

Vacuoles & Cytoplasmic Inclusions

Vacuoles

- Vaculoes are cavities in the cytoplasm and contain a fluid called cell sap.
- As the cell approaches maturity, some of the water-soluble reserve food materials manufactured by the cell get dissolved in the sap.
- Insoluble constituents precipitate out as cytoplasmic inclusion bodies.

Cytoplasmic Inclusions

- **Volutin**
  - are metachromatic granules.
  - Found to localize in the vacuoles of mature forms.
  - Contain inorganic polyphosphate, lipoprotein, RNA and magnesium.
  - May serve as phosphate storage structures.
- **Glycogen**

  Acumulates in the cytoplasm at the ends of the cell in the form of granules.
- **Fat globules**

  Found in the cytoplasm of bacteria in the form of fat globules.

Flagella & Endospores

Flagella

- Useful for motility of the bacteria.
- Bacteria belonging to Eubacteriales have peritrichous flagellus at all the sides.
- Some bacterial cells have polar flagella - a single flagellum or a tuft of flagella at one end.

Endospores

- Some bacteria transform themselves into small ovals or spheres which are highly resistant to adverse conditions.
- They are called as spores.

- They are produced intracellularly.
- Bacterial spores have dipicolinic acid which is absent in vegatative cells.

Fimbriae or Pill

Fimbriae

- Hair-like structures
- Observed as surface appendages.
- Present on some plant pathogenic bacteria.

Pill

- Hairlike structures found in some bacteria.
- They serve as adsorption organs for bacteriophages.
- Pili are considered to be sex organs.
- They mediate conjugation (mating) of bacteria.

Protoplast, Sphaeroplast & L forms

- All are bacterial forms.
- They devoid of the rigid cell wall.
- Bacterial cell lacking cell wall is called protoplast.
- Naked protoplast may synthesize a portion of cell wall material. Such a protoplast is called sphaeroplast.
- L forms are non-rigid bacterial forms. They are soft protoplasmic elements without defined morphology which can be propagated indefinitely on solid medium.

Chromosomes

- Chromosomes consist of proteins and nucleic acids.
- Chromosome maps show the location of various genes.
- Chromosomes an located in the chromatin body of the bacteria.
- They are the carriers of genes
  - Genes are factors that control inherited traits.
  - Genes are nucleic acids and more specifically DNA.
  - Genes are linked and aligned on the chromosome.

Plasmid & Episome

Plasmid

- Extrachromosomal DNA.
- Capable of autonomous replication.
- They are dispensible circular DNA strands.

- They are not capable of integration with the chromosome.
- Some bacteria have plasmids in the cytoplasm.

Episome

- Episomes are similar to plasmids.
- But unlike plasmids, episomes, can exist even integrated with the chromosome.
- They are also autonomous and dispensible genetic elements.
- Generally the bacteria containing plasmids do not have episomes and vice versa.

Transposon & Siderophore

Transposon

- Transposable genetic elements.
- Transpose = alter the positions of interchange.
- Mobile DNA segments that can insert into a few or several sites in a genome.
- Transponsons are even capable of moving between prokaryotes and eukaryotes.

Siderophore

- Produced by many bacteria extracellularly.
- They are low molecular weight iron (III) transport agents.
- Function of siderophores is to supply iron to the cell.
- Siderophores selectively complex with iron (III) (ferric iron) with very high affinity.
- Siderophore isolated from Pseudomonas sp. is called pseudobactin.

Bacteriocin

- Non-replicating, bactericidal protein-containing substances.
- They are produced by certain strains of bacteria and are active against some other strains of the same or closely related species.
- Bacteriocins include complete phage-like particles and small headed "killer particles" containing both protein and nucleic acid-as well as low molecular weight substances sensitive to proteases and nucleases.
- Bacteriocins are useful in the control of diseases.
- Bacteriocins produced by avirulent isolates are exploited to control diseases caused by virulent isolates of the same bacterium.
- Bacteriocins produced by Agrobacterium species are named as Agrocins.

- Bacteriocins produced by *Pseudomonas syringae* are called syringacins.
- Bacteriocin produced by *Agrobacterium radiobacter* strain 84 is called bacictiocin 84.
- High-molecular weight, trypsin-resistant bacteriocins are produced by strains of *Corynebacterium michiganense, Erwinia chrysanthemi, Pseudomonas phaseolicola, P. syringae, P. glycinea* and *P. solanacerum.*
- Low molecular weight bacteriocins are produced by some strains of *P. syringae, P. phaseolicola, P. glycinea* and *C. michiganense.*
- Particulate contractile bacteriocins are produced by *P. syringae 4-A* and *P. syringae W-1.*
- Killer particles are produced by strains 396 and 0362 of *Agrobacterium tumefaciens.*

Bacteriophage

- Viruses that infect bacteria are called bacteriophages.
- Bacteriophages are shaped like tadpoles with round, oval or polyhedral heads and a hollow tail with specialized fibrils at their tip.
- Head is composed of DNA surrounded by a protein membrane which covers the tail.
- Protein coat gives protection to the bacteriophage in its extracellular environment.
- Tail fibres attach to the cell wall of the bacteria.
- Tail fibres secrete an enzyme which penetrates the cell membrane to permit entry of the virus DNA.
- DNA enters through a pore and induces the bacterial cell to manufacture more virus particles.
- When bacteriophage particles multiply quickly, the bacterial cell dies.
- At the death of the cell, the bacteriophages are released & infect new bacterial cells.
- Bacteriophages survive only in bacterial cells and in the absence of them they perish.

6

Management of Bacterial Diseases of Plants

Control of Bacterial Diseases of Plants

1. Bacterial diseases of plants are usually very difficult to control.
2. Frequently, a combination of control measures is required to combat a given bacterial disease.
3. Using only healthy seeds or transplants.
4. Sanitation practices aiming at reducing the inoculums in a field.
5. Adjusting fertilizing and watering.
6. Crop rotation
7. Use of crop varieties resistant.
8. Resistant varieties, supplemented with proper cultural practices and chemical applications, are the most effective means of controlling bacterial diseases.

Soil and Seed Management

1. Soil infested with phytopathogenic bacteria can be sterilized with steam or electric heat and with chemicals such as FORMALDEHYDE, but this is practical only in greenhouses and in small beds or frames.
2. Seed, when infested superficially, can be disinfested with sodium hypochlorite or HCl solutions or by soaking it for several days in a weak solution of acetic acid.
3. If seeds can remain for 2 to 3 days in fermenting juices of fruit in which they are borne, bacterial pathogens can be eliminated.
4. When the pathogen is inside the seed coat and in the embryo, such treatments are ineffective.
5. Treating seed with hot water does not usually control bacterial diseases because of the relatively high thermal death point of the bacteria, but treatment at 52°C for 20 minutes often considerably reduces the number of infected seeds.

Chemicals control

1. Use of chemicals to control bacterial diseases has been, generally, much less successful than the chemical control of fungal diseases.
2. Of the chemicals used as foliar sprays, copper compounds give the best results.
3. Bordeaux mixture, fixed coppers, and cupric hydroxide are used most frequently for the control of bacterial leaf spots and blights.
4. Bacterial strains resistant to copper fungicides, however, are quite common.
5. Zineb, Maneb, or Mancozeb mixed with copper compounds is used for the same purpose, especially on young plants that may be injured by the copper compounds.

Antibiotics

1. Antibiotics have been used against certain bacterial diseases with mixed results.
2. Some antibiotics are absorbed by the plant and are distributed systemically.
3. They can be applied as sprays or as dips for transplants.
4. The most important antibacterial antibiotics in agriculture are formulations of STREPTOMYCIN or of STREPTOMYCIN AND OXYTETRACYCLINE.
5. Unfortunately, bacterial races resistant to antibiotics develop soon after widespread application of antibiotics; in addition, no antibiotics are permitted on edible plant produce.

Biological control

1. Successful practical biological control of the bacterial plant disease crown gall has been obtained by treating seeds or nursery stock with bacteriocin-producing antagonistic strains of *Agrobacterium.*
2. *Treatment of* tubers, seeds, and so on with antagonistic bacteria and spraying of aerial plant parts with bacteria antagonistic to the pathogen have given control of various diseases under experimental conditions but have been less successful in practice.

7

Pathogenicity Factors of Bacteria

Important events in pathogenesis of bacteria

- Recognition,
- Multiplication
- Translocation of Pathogen
- Stimulation or Inhibition of host

Determinants of Bacterial Pathogenicity

- Cell envelope components (capsule and extracellular polysaccharide (EPS))
- Flagella
- Pili
- Fimbriae
- Extracellular Enzymes
- Toxins
- Indole Acetic Acid (IAA).

Capsule & Flagella

Capsule

- Outermost component of the cell periphery.
- In *Erwinia amylovora*, non-capsular strains obtained as laboratory mutants or as naturally occurring strains are generally non-pathogenic.
- Non-capsular mutants of *Pseudomonas solanacearum* are avirulent.
- Capsule secretes extracellular polysaccharide (EPS).
- The amount of EPS production has been correlated with virulence of bacterial pathogens such as
    - *Xanthomonas campestris pv. oryzae*
    - *X. campestris pv. campestris*
    - *P. solanacearum*
    - *Erwinia stewarti*

Flagella

- Flagella help the bacteria for movement.
- Motile *Pseudomonas syringae pv. phaseolicola* causes more infection loci than paralyzed mutants.

Fimbriae & Enzymes

Fimbriae

- Fimbriae are commonly found in plant pathogenic bacteria.
- They function in adherence to surfaces of plant
- Example; *P. solanacearum, P. syringae pv. plastrobacterium tumefaciens and Erwinia carotovora.*

Enzymes

- Bacterial pathogens produce pectolytic enzymes which facilitate tissue disintegration.
- Example: *Erwinia carotovora* causing soft rot of vegetables.

Toxins &Indole acetic acid

Toxins

- *Pseudomonas syringae pv. phaseolicola* and *P. syringae pv. tabaci* are known to produce toxins.

Indole acetic acid

- IAA is produced by many plant pathogenic bacteria.
- It modifies the host metabolism and induces hypertrophy.
- Example: *Agrobacterium tumefaciens* produces IAA in infected tissues

8

Phytoplasmas & Spiroplasmas

1. In 1967, wall-less microorganisms were seen with the electron microscope in the phloem of plants infected with one of several yellows-type diseases and in insect vectors of these diseases.
2. New microorganisms were subsequently called mycoplasma like organisms because of their superficial resemblance to mycoplasmas.
3. It was later shown that these organisms are not mycoplasmas. Although all are mollicutes, i.e., prokaryotic cells without cross walls, a few of them have a helical structure and are called spiroplasmas.
4. Most, however, are round to elongate but are not spiral and are now called phytoplasmas.

Characteristics Symptoms

1. Gradual, uniform yellowing or reddening of the leaves.
2. Smaller leaves.
3. Shortening of the internodes and stunting of the plant.
4. Excessive proliferation of shoots.
5. Formation of Witches'-brooms.
6. Greening or sterility of flowers.
7. Reduced yields.
8. Finally, a more or less rapid dieback, decline, and death of the plant.
9. Root abnormalities and necrosis often precede the aboveground symptoms.

Properties of True Mycoplasmas

1. Prokaryotic organisms that have no cell walls.
2. They are members of the class Mollicutes, which has one order, Mycoplasmatales.
3. The order has three families, each with one genus:
   - Mycoplasmataceae, genus *Mycoplasma.*
   - *Acholeplasmataceae, genus Acholeplasma.*
   - Spiroplasmataceae, genus *Spiroplasma.*

4. They lack a true cell wall, mycoplasmas are bounded only by a "unit" membrane.
5. Shapes range from spherical or slightly ovoid to filamentous.
6. Reproduce by budding and by binary transverse fission of cells.
7. Have no flagella, produce no spores, and are gram negative.
8. Nearly all mycoplasmas parasitic to humans and animals.
9. Grown on more or less complex artificial nutrient media in which they produce minute colonies that usually have a characteristic "fried-egg" appearance.
10. Most mycoplasmas are completely resistant to penicillin; however, they are sensitive to tetracycline and chloramphenicol, and some are sensitive to erythromycin and to certain other antibiotics.

Phytoplasmas

1. Organisms observed in plants and insect vectors, i.e., the phytoplasmas, which do not include the spiroplasmas, resemble mycoplasmas of the genera Mycoplasma or Acholeplasma in all morphological aspects.
2. Genetically, phytoplasmas are more related to Acholeplasma than to Mycoplasma.
3. They lack cell walls, are bounded by a "unit" membrane, and have cytoplasm, ribosomes, and strands of nuclear material.
4. The size of their chromosomes varies from 530 kilobases of DNA to 1130 kilobases.
5. Their shape is usually spheroidal to ovoid or irregularly tubular to filamentous.

Transmission

1. Phytoplasmas (and spiroplasmas) are generally present in the sap of a small number of phloem sieve tubes.
2. Most plant mollicutes are transmitted from plant to plant by leafhoppers but some are transmitted by psyllids and plant hoppers.
3. Plant mollicutes also grow in the alimentary canal, hemolymph, salivary glands, and intracellularly in various body organs of their insect vectors.
4. Vector cannot transmit the mollicutes immediately after feeding on the infected plant.
5. It begins to transmit them after an incubation period of 10 to 45 days, depending on the temperature.
6. Shortest incubation period occurs at about 30°C, the longest at about 10°C.

7. Mollicutes can be acquired as readily or better by nymphs than by adult leafhoppers and survive through subsequent molts.
8. They are not passed from the adults to the eggs and to the next generation.
9. Therefore, must feed on infected plants in order to become infective vectors.

Sequence of events in the Overwintering, Acquisition & Transmission

Diagnosis

1. Phytoplasmas has not yet been cultured on artificial media.
2. Specific antibodies, DNA probes, RFLP profiles, and analysis of 16 S rRNA genes with the help of the PCR amplification technique have become extremely useful in the detection and identification of the pathogen in suspected hosts.
3. Earlier methods included indexing to sensitive hosts, fluorescent staining with either the DNA-specific stain 4,6-diamidino-2-phenylindole (DAPI) or the callose-specific stain aniline blue, or staining with the so-called Dienes' stain.

Management

1. Mollicutes are sensitive to antibiotics, particularly those of the tetracycline group.
2. When infected plants are immersed in or injected with tetracycline solutions, the symptoms, if already present, recede or disappear or, if not yet present, are delayed.
3. Foliar and soil application are ineffective.
4. Infected plants or dormant propagative organs can be totally freed of mollicutes by heat treatment.
5. Infected plants are kept in growth chambers at 30 to 37°C for several days, weeks, or months; dormant organs are immersed in hot water at 30 to 50°C for as short as 10 minutes at the higher temperatures and as long as 72 hours at the lower temperatures.

Spiroplasmas

1. Spiroplasmas are cells that vary in shape from spherical or slightly ovoid.
2. Unlike phytoplasmas, spiroplasms can be obtained from their host plants or their insect vectors and cultured on nutrient media.
3. They produce mostly helical forms in liquid media.
4. They lack a true cell wall and are bounded by a unit membrane.

5. They multiply by fission.
6. The helical filaments are motile.
7. There are no flagella.
8. Colonies of spiroplasmas on agar have a diameter of about 0.2 millimeters; some have a typical fried-egg appearance, but others are granular.
9. Spiroplasmas are resistant to penicillin but are inhibited by tetracycline.
10. Cultured plant spiroplasmas can be injected into or fed to their insect vectors, which then, on feeding on the host plants, transmit the organisms to the plants.
11. The experimentally infected hosts develop typical symptoms of the disease.
12. Spiroplasmas are known to cause the stubborn disease in citrus plants and the brittle root disease in horseradish (*Spiroplasma citri*), stunt disease in corn plants, and a disease in periwinkle.
13. Spiroplasmas are helical mollicutes.
14. *Spiroplasma citri* has also been found in many other dicots, such as crucifers, lettuce, and peach, and both *S. citri* and the corn stunt spiroplasma also infect their respective leafhopper vectors.
15. Several kinds of spiroplasmas have been shown to infect honeybees and several other insects.
16. Several more live saprophytically on flowers and other plant surfaces and, possibly, internally in plants.

Unit 6: Plant Virology

1

Nature, Composition and Architecture of Viruses

Characteristic of Virus

- Submicroscopic.
- Infectious obligate pathogen.
- Do not have metabolism of their own but depend upon other living cells for their multiplication.
- Viruses contain a single type of nucleic acid, either RNA or DNA, never both.
- Lipman system is absent in viruses.
- They neither have any sexual reproduction nor produce any resting structures as that of fungi and bacteria.

Nature of Virus

- Each plant virus consists of two components the nucleic acid and the protein coat or capsid.
- Mature particle of a plant virus is generally called virion.
- Whole infective particle is called as Nucleocapsid.

Nucleic Acid of Viruses

- Nucleic acid portion of the virus particle is called as Genome.
- Genomes are organized as single nucleic acid molecules that are linear or circular.
- They may have as few as four genes or as many as several hundred.
- They may be double-stranded DNA, single-stranded DNA, double-stranded RNA, or single-stranded RNA.
- Genome is actual infective component.
- Majority of plant viruses contain RNA.
- Some viruses (Cauliflower mosaic virus, Maize streak virus, Mung bean yellow mosaic virus) contain DNA.

- Most plant viruses contain single strand of RNA.
- In double stranded viruses (as RNA or ds DNA) the two strands are coiled around each other helically.

Virus Protein

- Protein coat that encloses the viral genome is called capsid.
- Its structure may be rod-shaped, polyhedral, or complex.
- Composed of many capsomeres, protein subunits made from only one or a few types of protein.
- It helps viruses infect their host.
- It protects nucleic acid (RNA or DNA) of virus.
- It is made-up of different amino acid sequences in different viruses.
- Protein shell or coat protects viral nucleic acid from environment.

Chemical Composition of a Virus

- Plant viruses contain selective amounts of nucleic acids and protein.
- Viruses with isometric particles may contain between 15% and 45% nucleic acid.
- Viruses with rod shaped particles have only about 5% nucleic acid.
- They also contain polyamines, enzymes and lipids.

Do virus have life?

Viruses are non-living

- Viruses have no cellular structure.
- Viruses do not respire and show no metabolism.
- They do not reproduce sexually/asexually.
- They can be crystallized and still retain the infectivity.
- They possess high specific gravity.

Viruses are living

- Strictly obligate parasites.
- Respond to stimuli like heat and light.
- Multiply enormously.
- Undergo incubation period.
- Show properties of mutation.
- Infective and capable of spreading disease.

Tulipomania late 16th century

- Known to be caused by a virus.
- Tulips with color breaking symptoms were prized and traded for large sums of goods- this led to "tulipomania" in the late 1500's.
- Traded for 1 Viceroy tulip bulb: 4 tons of wheat,8 tons of rye,4 barrels of beer.

Different types of Viruses

Sl.No	Types		Examples
1.	**RNA viruses**		
	a. Single stranded	With one segment	• Tobacco mosaic viruses • Lettuce necrotic yellow virus
		With two segments	• Cowpea mosaic virus • Tobacco ring spot virus
		With three segments	• Cucumber mosaic virus
		With more number of segments	• Potato virus x • Tomato spotted wilt virus
	b. Double stranded		• Wound tumour virus • Fiji disease virus • Reo virus
2.	**DNA viruses**		
	Double stranded		Cauliflower mosaic virus
	Single stranded		Maize streak virus, ϕX174

VIRION

- The individual, complete and infectious nucleoprotein particles of a virus.
- In short, virus particles are called virions.
- They are also called nucleocapsids.
- Protein subunits form the protein coat, shell or capsid and they are called structural units.
- Clusters of such units are called morphological units or capsomers.
- Envelope is called as peplos.

SATELLITE VIRUS

- Satellite virus has a coat of its own but a small RNA genome.
- It needs another virus to supply replicase and probably other enzymes.
- It acts as a parasite of plant-parasitic virus.
- One example is tobacco necrosis satellite virus.

SATELLITE RNA

- Satellite RNA has no coat protein.
- It is associated with particles of some viruses.
- It completely depends upon those viruses for multiplication.
- Viruses which help multiplication of satellite RNA are called helper viruses.
- Satellite RNA aggravates or reduces the symptom development induced by the helper virus.
- This type of association has been reported in cucumber mosaic virus and tobacco ring spot virus.

Satellite RNA depending upon Satellite Virus

- Satellite tobacco necrosis virus has been shown to contain a small satellite RNA.
- That is dependent on tobacco necrosis virus for replication and on the satellite virus for encapsidation.

Viroid

- Viroids are naked nucleic acids without coat protein.
- They consist of only RNA.
- These "miniviruses' are the smallest known agents of infectious disease.
- They are subviral in size.
- Some of the viroids are: Citrus excocortis viroid, potato spindle tuber viroid, chrysanthemum stunt and chlorotic mottle viroid, cucumber pale fruit viroid and hop stunt viroid.

Virusoid

- Some viruses contain a viroid-like satellite RNA in addition to a linear single-stranded molecule of genomic RNA. They are called virusoids.
- They show little sequence homology with the viroids.
- But they do show significant homology with a linear satellite RNA associated with tobacco ring spot virus.

Multipartite virus

- Some viruses have different particle types which differ in size or nucleic acids.
- In some mutipartite viruses, light and heavy particles exist.
- Only heavy particles are infectious; eg. Turnip yellow mosaic virus.
- But in general, all the types of virus particles are simultaneously required for successful infection. e.g. Tobacco rattle virus.

Helper viruses

- Viruses which help in multiplication of satellite RNA are called helper viruses.

Temperate viruses

- Viruses that can integrate their genome into a host chromosome and remain latent until they initiate a lytic cycle.

Provirus

- Viral DNA that inserts into a host cell chromosome.

Retrovirus (Retro = backward)

- RNA virus that uses reverse transcriptase to transcribe DNA from the viral RNA genome.

Morphology/Architecture of Viruses

1. Elongated (Rigid Rod or Flexious thread)

Sl.No	Morphology	Example	Size(nm)
1	Rigid rod	*Barley stripe mosaic virus*	20X 10
		Tobacco mosaic virus	15 X 300
2	Flexuous thread	*Potato virus X*	10-13 X480
		Citrus tristeza virus	10-14 X 2000

2. Spherical (Isometric or Polyhedral)

All spherical viruses are actually polyhedral ranging in diameter about 17 nm to 60 nm. Examples;

Sl.No	Example	Diameter (nm)
1	Tobacco Necrosis satellite virus	17
2	Wound tumour virus	60
3	Tomato spotted wilt virus	70-80

MORPHOLOGY

Shape and size

- Plant viruses are smaller than bacteria.
- Seen only with electron microscope.
- Measured in nanometers or millimicrons or millimues (1 nm = 0.001µm or 1 x 10 mm or 1 x 10-9 mt).

Plant viruses have different shapes and sizes

Helical symmetry

Nearly half of the known viruses falls under category of elongated rods

Rod-shaped virus

- About 20-25 nm in diameter and from about 100 to 300 nm long.
- These appear rigid and often have a clear central canal.

Plant Rhabdovirus

- Also known as bullet shaped or Bacilliform.
- Like a bullet they have both or one end rounded and the ratio between their length and breadth is very small.
- Example: Cocoa swollen shoot virus.
- Genus: Badnavirus-28 x 130 nm.

Isometric

- Spherical virus may be isometric or polyhedral (17 nm - 60 nm dia) in shape.
- Example: Tobacco necrosis satellite virus.

Icosahedral

- Also known as spherical, quasi spherical, polyhedral or icosahedral.
- Polyhedral (many sided).
- Icosahedral (20 sided polyhedrons, 12 corners).
- All sides are of equal measurements and all faces.
- Ex.. triangles.

Multipartite Plant Viruses

Segmented virus

- Genomes are those which are divided into two or more physically separate molecules of nucleic acid.
- All of which are then packaged into a single virus particle.

Multipartite genomes

- Segmented, but each genome segment is packaged into a separate virus particle.
- Geminivirus (single-stranded DNA) Bipartite
- Comovirus (single-stranded RNA) Bipartite
- Furovirus (single-stranded RNA) Bipartite
- Tobravirus (single-stranded RNA) Bipartite
- Partitiviridae (double-stranded RNA) Bipartite
- Bromoviridae (single-stranded RNA) Tripartite
- Hordeivirus (single-stranded RNA) Tripartite

2

Properties of Plant Viruses

Viral Protein

1. Provides a protective sheathing for the nucleic acid of the virus.
2. Plays a role in determining vector transmissibility of a virus.
3. Play the kinds of symptoms it causes.
4. Protein itself has no infectivity, but its presence generally increases the infectivity of the nucleic acid.
5. Some proteins have been shown to facilitate the movement of the virus through cells.
6. Others to be required for transmission of the virus by its vector.
7. Some for production of proteins needed for cleaving the nucleic acid of the virus in precise positions.
8. Some for producing the cellular inclusion bodies observed in cells infected by viruses.
9. Diseased condition induced in plants by viruses is the result of the interference and disruption of normal metabolic processes in infected parenchyma or specialized cells.
10. Such interference is caused by the mere presence and multiplication of the virus and, possibly, by the abnormal or toxic effects of additional virus-induced proteins or their products, although no such substances have been found to date.

Genomic Nucleic acid

1. Infectivity of viruses is strictly the property of their genomic nucleic acid, which in most plant viruses is RNA.
2. Some viruses carry within them a transcriptase enzyme that they need in order to multiply and infect.
3. Capability of the viral RNA to reproduce both itself and its specific protein indicates that the RNA carries all the genetic determinants of the viral characteristics.

4. Expression of each inherited characteristic depends on the sequence of nucleotides within a certain area (gene) of the viral RNA, which determines the sequence of amino acids in a particular protein, either structural or enzyme.
5. This is called coding and seems to be identical in all living organisms and the viruses.
6. The code consists of coding units called codons.
7. Each codon consists of three adjacent nucleotides and determines the position of a given amino acid in the protein being synthesized.
8. The amount of RNA, then, contained in each virus indicates the approximate length of, and the number of nucleotides in, the viral RNA.
9. This in turn determines the number of codons in each RNA and, therefore, the number of amino acids that can be coded for.
10. In some viruses, the amount of nucleic acid available for coding is increased by having some genes overlap parts of or whole other genes, or by frame shifting, i.e., reading the nucleotides in a different sequence from the first one and thereby forming entirely different codons and genes.
11. Because the protein subunit of viruses contains relatively few amino acids (158 in TMV), the number of codons utilized for its synthesis is only a fraction of the total number of codons available (158 of 2,130 in TMV).
12. In addition to protecting the viral nucleic acid, the coat protein in some cases affects, as mentioned already, the symptoms caused by the virus, the movement of some viruses in their hosts, and transmission of viruses by their vectors.
13. The remaining codons are presumably involved in the synthesis of other proteins, either structural proteins or enzymes.
14. One of these enzymes is called an RNA polymerase (RNA synthetase or RNA replicase) and is needed to replicate the RNA of the virus.

Diseases Caused by Negative RNA; [(-) SSRNA] Viruses

1. Negative RNA plant viruses include those belonging to the families Rhabdoviridae and Bunyaviridae.
2. Both of which also contain viruses that infect humans and animals, and to the genera *Tenuivirus and Ophiovirus,* which have not yet been assigned to families.
3. All negative RNA viruses carry a transcriptase enzyme (RNAdependent RNA polymerase) within their particle that transcribes the (-) ssRNA to (+) ssRNA so it can be translated.

3

Nomenclature and Classification of Viruses

- Several plant viruses are named after the most prominent symptom they cause on the first host.
- A virus causing a mosaic on tobacco is called tobacco mosaic virus.
- Disease itself is called tobacco mosaic.
- Virus causing spotted wilt symptoms on tomato is called tomato spotted wilt virus and the disease is called tomato spotted wilt.
- All viruses belong to the kingdom Viruses.
- Within the kingdom, viruses are distinguished as RNA viruses and DNA viruses.
- It depends on whether the nucleic acid of the virus is RNA or DNA.
- Viruses are further subdivided depending on whether they possess one or two strands of RNA or DNA of either positive or negative sense, either filamentous or isometric.
- Within each of these groups there may be viruses replicating via a polymerase enzyme (+RNA or DNA viruses) or via a reverse transcriptase (-RNA or DNA viruses).
- Most viruses consist of nucleic acid surrounded by coat protein, but some also have a membrane attached to them.
- Some viruses have all their genome in one particle (monopartite viruses), but the genome of other viruses is divided among two, three, or, rarely, four particles (multipartite viruses).
- Other characteristics in the classification of viruses include
  - Symmetry of helix in the helical viruses.
  - Number and arrangement of protein subunits in the isometric viruses.
  - Size of the virus.
  - Finally, any other physical, chemical, or biological properties.

Classification of Viruses

	Name		Pieces of RNA/DNA	Shape
Single-stranded positive RNA [(+) ssRNA				
Rod-shaped particles	**Genus**	**Type species**	**Remarks**	
1 ssRNA	Tobamovirus	Tobacco mosaic virus	Contact transmission	
2 ssRNAs	Tobravirus	Tobacco rattle virus	Nematode transmission	
3 ssRNAs	Furovirus	Soilborne wheat mosaic	Fungal transmission	
3 ssRNAs	Pomovirus	Potato mop-top virus	Fungal transmission, Dicot	
4 ssRNAs	Benyvirus	Beet necrotic yellow vein virus	Fungal transmission	
Filamentous particles				
1 ssRNA	Allexivirus	Shallot virus X	Eriophyid mite transmission	
	Foveavirus	Apple stem pitting virus	No vector	
	Potexvirus	Potato virus X	By contact only	
	Capillvirus	Apple stem grooving virus	No vector. Some seed transmission	
	Trichovirus	Apple chlorotic leaf spot virus	No vector. Some seed transmission	
	Vitivirus	Grapevine virus	Mealybugs, Scale insects, Aphids	

Group			
Single-stranded positive RNA [(+) ssRNA			
Isometric particles	**Genus**	**Type species**	**Remarks**
1 ssRNA	Sobemovirus	Southern bean mosaic virus	Seedborne, Beetles, Myrids
	Marafivirus	Maize rayado fino virus	In Gramineae, leafhoppers
	Umbravirus	Carrot mottle virus	Aphids / Helper virus
	Tymovirus	Turnip yellow mosaic virus	By beetles
2 ssRNAs	Ourmiavirus	Ourmia melon virus	No vectors known
Bacilliform particles			
3 ssRNAs	Ourmiavirus	Ourmia melon virus	No vectors known
Filamentous viruses			
1 ssRNA	Potyvirus	Potyvirus	Aphids, / helper virus
	Ipomovirus	Sweet pot mild mottle	Whitefly Bemisia tabaci

	Rymovirus	Ryegrass mosaic virus	Eriophyid mites
	Bymovirus	Barley yellow mosaic virus	Gramineae, fungal transmission
1 or 2 ssRNA	Closterovirus	Beet yellows virus	Aphids, mealybugs or whiteflies
	Crinivirus	Lettuce inf. yellows virus	Whiteflies
Isometric viruses			
1 ss(+)RNA	Waikavirus	Rice tungro spherical virus	Leafhoppers or aphid
	Machlomovirus	Maize chlorotic mottle virus	Seed, beetles, thrips
	Necrovirus	Tobacco necrosis virus	Fungal transmission
	Luteovirus	Barley yellow dwarf virus	Luteovirus Barley yellow dwarf virus Gramineae, aphids
	Polerovirus	Potato leafroll virus	Monocot or dicot plants
	Enamovirus	Potato leafroll virus	Monocot or dicot plants
2 ss(+)RNA	Comovirus	Cowpea mosaic virus	Chrysomelid beetles
	Nepovirus	Tobacco ring spot virus	Nematodes
	Fabavirus	Broad bean wilt virus	Aphids
3 ss(+)RNA	Cucumovirus	Cucumber mosaic virus	Aphids
	Alfamovirus	Alfalfa mosaic virus	Aphids
	Alfamovirus	Tobacco streak virus	Pollen, seed
dsRNA			
	Varicosa virus	Lettuce big-vein virus	Fungal transmission
	Varicosa virus	White clover crypto. virus 2	Same, seed
	Alphacrypto virus	White clover crypto. virus 1	Nonenveloped, latent
	Oryza virus	Rice ragged stunt virus	Planthoppers
	Fijivirus	Fiji disease virus	Gramineae, planthoppers
	Phytoreo virus	Wound tumor virus	Leafhoppers
(-) ssRNA			
Bacilliform particles	Cytorhabdo virus	Lettuce necrosis	Leafhoppers, yellows virus, Planthoppers, Aphids
	Nucleorhabdo virus	Potato yellow dwarf virus	Same

Membranous circular particles	Tospovirus	Tomato spotted wilt virus	Thrips
Thin flexuous multipartite viruses	Tenuivirus	Rice stripe virus	Gramineae, planthoppers
ds DNA			
Isometric	Caulimovirus	Cauliflower mosaic virus	Aphids
		Soybean chlorotic mottle virus-like	Aphids
		Cassava vein mosaic virus-like	Aphids
		Petunia vein clearing virus-like	Aphids
		Rice tungro bacilliform virus-like	Leafhoppers
	Badnavirus		Mealybugs, Leafhoppers
ds DNA			
Isometric	Caulimovirus	Cauliflower mosaic virus	Aphids
		Soybean chlorotic mottle virus-like	Aphids
		Cassava vein mosaic virus-like	Aphids
		Petunia vein clearing virus-like	Aphids
		Rice tungro bacilliform virus-like	Leafhoppers
	Badnavirus	Commelina yellow Mealybugs mottle virus	Mealybugs
(+)ssDNA			
	Mastrevirus	Maize streak virus	Gramineae, leafhoppers
	Curtovirus	Beet curly top virus	Dicot, leafhoppers
	Begomovirus	Bean golden mosaic virus	2 DNAs, whiteflies
	Topocuvirus	Tom pseudocurly top vrius	Treehopper
	Nanovirus	Subteranean clover stunt virus	6 DNAs

4

Variability in Viruses

- Virus show variability.
- Variability in virus is due to
  1. Recombination
  2. Mutation
  3. Selection pressure
  4. Pseudo recombination

Variability due to Mutation

- Mutation is also a important source of origin of mutants.
- Mutations can be base substitutions and deletion mutations.
- Base substitution results in point mutation whereby base of a nucleotide is changed so that encoded protein molecule of parent strain.
- Such base substitutions have been discovered in cauliflower mosaic viruses and tobacco mosaic.
- Deletion mutations occur when part of the genome is deleted so that the mutant has lost the character encoded by the deleted segment of genome.
- Deletion mutations were first reported by Reddy and Black (1974, 1977) in wound tumor virus.
- Since the deletion mutations have been discovered in several other viruses.

Variability due to Recombination

- Plant viruses undergo recombination.
- Two strains of the same virus are inoculated into the same host plants.
- One or more new virus strains are recovered with properties (symptomatology, virulence etc) different from those of either of the original strains combinants.
- For example. two tobravirus strains called N_5 and 16 are formed by natural recombination of $RNAs_2$ of tobacco rattle and pea early browning viruses.

- Formation of new strains also occurs in brome mosaic virus by genetic recombination.

Variability due to Pseudo Recombination

- RNA genome of multicomponent viruses exists as more than one segments.
- RNA segments of closely related plant viruses can be exchanged to create new viable strains, called pseudorecombinants and the process as pseudorecombination.
- Cucumber mosaic virus Q strain is a pseudorecombinant.

Variability due to Selection Pressure

- Selection pressure resulting in the selection of new strains under changed conditions.
- With a change in selection pressure, an earlier minor strain of the mixture becomes the new dominant strain which seemingly leads to origin of new strains.
- Temperature and host are the two most common and important selection pressure operative in nature leading to the selection of new strains.

5

Satellite Viruses and Satellite RNAs

Typical viruses consist of nucleic acid contained in a capsid composed of protein molecules that can multiply and cause infection by themselves.

In addition to typical viruses, however, two other types of virus-like pathogens are associated with plant diseases. Satellite Viruses and Satellite RNAs.

Satellite viruses

1. Are viruses but cannot cause infection by themselves.
2. They must always be associated with certain typical viruses (helper viruses) because they depend on the latter for multiplication and plant infection.
3. Satellite viruses often reduce the ability of the helper viruses to multiply and cause disease; i.e., satellite viruses act like parasites of the associated helper virus.

Satellite RNAs

1. Small, linear or circular RNAs found inside virions of certain multicomponent viruses.
2. Not related, or are only partially related, to the RNA of the virus.
3. Satellite RNAs may increase or decrease the severity of viral infections.

6

Conventional & Biotechnological Techniques Used in Detection & Diagnosis of Viruses

Detection/Diagnosis

Detection

- An art or skill of identifying some abnormalities, i.e. which is below or higher than the normal permissible range/threshold.
- Detection means the existence of whatever do you want detecting but diagnosis is that the analysis in depth of existing thing.
- Objectively look for, listen to the credible complains, symptoms that is sometimes not visible at all.

Diagnosis

- To ascertain the specific medical condition that is causing the problem.
- It a complex problem, where based on known theory, physiology, pathological findings, his experience one has to come to conclusion and answer to the question that "Why such symptoms exist? "
- It is a-priory step for prescribing medications for the plants.

Detection and Identification of Plant Viruses

Because viruses are too small to be detected with the naked eye or seen through a light microscope, their presence has been detected primarily by

- Symptoms exhibited by the host plant.
- Symptoms induced in an indicator plant after transmission of the virus by grafting.
- Mechanical inoculation.
- by one of the vectors.
- by examination of young infected tissues with a microscope for cell inclusion bodies that may be characteristic of the virus family or genus;
- by electron microscopy.

- by one of the serological tests such as ELISA or fluorescent antibody microscopy.
- by isolating, and subsequently analyzing through electrophoresis, the dsRNA of viruses replicating in plants because healthy plants do not produce such dsRNAs.
- Inclusion body detection technique, has the advantage that it can be used to detect known as well as unknown viruses for which no antiserum or not much information is available, and it can therefore be used for detection of even woody plant viruses.

Conventional Techniques used in Detection and Diagnosis of Viruses

1. Once the cause of a disease has been established as a virus, a series of tests, utilizing whatever simple or sophisticated methods and equipment are available, may be necessary to determine its identity.
2. Host range of the virus, i.e., the hosts on which the virus induces symptoms and the kinds of symptoms produced, may help to differentiate the virus from several others.
3. Transmission studies should indicate whether the virus is transmitted mechanically and to what hosts, or by insects and which insects and so on, with each new property ascertained helping to characterize the virus further.
4. If the virus is transmitted mechanically, certain properties of the virus, such as its. Thermal Inactivation Point: *the temperature required for complete inactivation of the virus in untreated crude juice during a 10-min exposure*), its longevity *in vitro*, and its Dilution End Point (i.e., *the highest dilution of the juice at which the virus can still cause infection*), have been used in the past but are not reliable.
5. Serological Tests: Used if the identity of the virus is suspected and if they are positive, a tentative identification may be made.
6. Electron Microscope: Examination of the virus in an Electron Microscope and inoculation of certain plant species are also usually sufficient for a tentative identification of the virus.
7. **Indexing**: In virus-like diseases of woody (and other) plants in which no pathogens have been observed so far, identification of the pathogens, which are at present presumed to be viruses, is made strictly by **indexing**.
8. **Indicators**: Indexing involves inoculation by grafting of certain plant species or varieties called **indicators**.

- Indicators are sensitive to specific viruses and on inoculation with these viruses develop characteristic symptoms and vice versa.
- Development of the characteristic symptoms by an indicator identifies the virus with which the indicator was inoculated.

SEROLOGY

- Alien proteins like that of virus particles can induce the production of specific antibodies in certain cells of lymphatic tissues in animals.
- The alien protein molecules act as antigens.
- When cleared blood serum, called antiserum is mixed with the virus suspension precipitates are formed due to specific binding of the virus protein (antigen) with antibodies.
- The precipitates are visible to the naked eye.
- The study of such reactions is called serology.

Use of Serological techniques

1. To determine relationships between viruses.
2. To identify a virus causing causing plant disease.
3. To detect virus in foundation stocks of plants.
4. To detect symptomless viral infection.
5. To measure virus quantitatively.
6. To locate the virus within a cell or tissue.
7. To detect plant viruses in insects and to purify a virus.

Antiserum

- Antisera are prepared by injecting purified virus suspension or suspensions of virus protein intravenously, intramuscularly or both into rabbits or guinea pigs or mice.
- Normal plant proteins found in the virus suspensions may also induce antibodies.
- They are removed by saturating the, antiserum with sap from healthy plants and then removing the precipitates by centrifuging.
- The clear antiserum can, be stored in refrigerator or deep-freeze or in sealed ampoules after freeze-drying.
- Antisera have been prepared against a wide range of plant viruses.
- These antisera are highly useful for detecting virus particles and virus infections for the diagnosis of virus diseases.
- Several sensitive serology tests have been developed to detect virus infection.

Monoclonal Vs Polyclonal antibodies

Antiserum: Serum containing antibodies

Monoclonal

- Antiserum contain single antibody.
- Each antibody react only with single antigen of pathogen.

Polyclonal

- Antiserum contains a mixture of different antibodies.
- Each antibody reacting with the antigen but at a different surface locality.

Serology Tests to Detect Virus Infection

Precipitation Test

- Clarified virus suspension (antigen) and antiserum (antibody) are mixed in narrow glass tubes.
- Precipitates are formed when antigen and antibodies are related.

Ouchterlony Test

- Wells are cut in the agar on the bottom of a petri dish.
- Antiserum (antibody) is placed on the central well while the virus suspension (antigen) is placed on surrounding wells.
- Antibodies and antigens move towards each other and bands are formed when they meet.
- The test is also called double diffusion test.

Chloroplast-Agglutination Test

- Crude sap with the virus is mixed with antiserum on a microscope slide.
- Chloroplasts clump together with virus in this reaction.

Latex Test

- Polystrene latex spheres are first sensitise with antibodies.
- Sensitized latex can be stored at 4 for years.
- By mixing the latex with virus suspension, precipitation is obtained.

ELISA (Enzyme linked Immuno-sorbent assay)

- Antibody is first linked with an enzyme like alkaline phosphatase.
- Virus is then mixed with the antibody enzyme.
- The substrate (p-nitrophenyl phosphate for alkaline phosphatase) is degraded by the enzyme action and bright yellow colour appears.
- The colouration can be measured in a spectrometer.

- The reaction will be positive only when the virus (antigen) is related to the antibody.
- The test is extremely sensitive and useful is detecting viruses that occur in low concentration.

Biotechnological Techniques used in Detection and Diagnosis of Viruses

Molecular Technique

Any properties of an individual's showing heritable variations is referred to as a character.

These characters which can be easily identified are referred to as marker.

Types of markers

Sl.No	Name of marker	Remarks
1.	Morphological Markers	• Markers related to size, color, and surface of the various parts. • Such characters are used for varietal identifications
2.	Biochemical Markers	• Markers related to variations in protein and amino acids banding patterns. • Gel electrophoresis studies are used for identification of biochemical markers
3.	Molecular markers	• A unique sequence of nucleotides found on a strand of DNA. • Markers related to variations in DNA fragments. generated by restriction endonuclease enzymes. • DNA markers are also used as molecular markers. • Also called genetic markers or DNA Markers.

Polymerase Chain Reaction (PCR)

1. Also called molecular photocopying.
2. PCR technique was invented by Kary B.Mullis in 1985.
3. It is a molecular biology technique for enzymatically replicating DNA without using living organisms like *E.coli*, Yeast.
4. PCR are used to amplify or make many copies of small fragment of DNA.
5. PCR process is directed by a machine called thermocycler.
6. The machine is programmed to alter the temperature of the reaction every few minutes to allow DNA denaturing and synthesis.
7. Taq polymerase enzyme, which is tolerant to high temperature is used in the PCR process.

8. Taq polymerase was discovered in hot springs from the bacteria *Thermus auquaticus*.
9. DNA polymerase is stable at high temperatures needed to perform the amplification.

Procedure of PCR

Each cycle of PCR consists of three important steps

Denaturing

1. Achieved at high temperature (94-96^{0}C).
2. Makes both templates DNA primers single stranded.
3. Takes usually 1-2 minutes.

Annealing

1. Achieved by lowering the temperature (45-60^{0}C).
2. Leads to attachment of primers (single stranded) to single DNA strands.

Elongation

1. Allows DNA polymerase to copy to DNA strand.
2. Final elongation step is used after the last cycle.

Biotechnological Techniques Used in Detection and Diagnosis

1. Isolated single- or double-stranded RNA can be used further to produce complementary DNA (cDNA) to the RNA. cDNA can also be produced to single- or double stranded DNA.
2. The cDNA, if produced in the presence of radioactive or chromogenic (color producing) molecules, can be used for hybridization experiments with viral RNA or DNA.
3. The viral RNA or DNA is partially purified from suspected infected plants and is allowed to react with cDNA in test tubes or on nitrocellulose filter supports (Dot blots).
4. The cDNA–RNA or cDNA–DNA hybrids are detected and quantified by autoradiography or liquid scintillation counting, if radioactive, or by colorimetric techniques, if chromogenic.
5. In addition to virus detection via the formation of cDNA–RNA hybrids, cDNA to the virus RNA can be further converted to dsDNA, which can then be cloned into suitable vectors (e.g., *Escherichia coli* bacteria) to produce almost unlimited amounts of dsDNA and, from this, cDNA probes for further hybridization experiments for virus detection.

6. The PCR technique allows unlimited amplification of selected specific DNA sequences for which suitable primers (short DNA sequences) are available.
7. Combining reverse transcription of viral RNA into DNA allows use of PCR amplification for RNA as well as DNA viruses.
8. Once amplification of the nucleic acid is accomplished, use of labeled DNA probes or electrophoretic analysis of the PCR products allows further detection and diagnosis of the virus.

7

Behaviour of Viruses in Plants Including Infection, Replication and Movement

Virus Infection and Virus Synthesis

Plant viruses enter cells only through WOUNDS made mechanically or by VECTORS or by deposition into an ovule by an INFECTED POLLEN GRAIN.

I. Replication of an RNA virus

- RNA of the virus is first freed from the protein coat.
- Then induces the cell to form the viral RNA polymerase.
- This enzyme utilizes the viral RNA as a template and forms complementary RNA.
    - The first new RNA produced is not the viral RNA but a mirror image
    1. As the complementary RNA is formed, it is temporarily connected to the viral strand.
    2. Thus, the two form a double-stranded RNA that soon separates to produce the original virus RNA and the mirror image (-) strand, with the latter then serving as a template for more virus (+strand) RNA synthesis.
- In single-stranded RNA rhabdoviruses, the RNA is not infectious because it is the (-) strand.
- This RNA must be transcribed by a virus-carried enzyme called transcriptase into a (+) strand RNA in the host, and the latter RNA then replicates as described earlier.
- In double-stranded RNA isometric viruses, the RNA is segmented within the same virus, is noninfectious, and depends for its replication in the host on a transcriptase enzyme also carried within the virus.

II. Replication of double-stranded DNA virus

1. On infection of a plant with a double-stranded DNA (dsDNA) virus, the viral dsDNA enters the cell nucleus, where it appears to become twisted and supercoiled and forms a minichromosome

2. The latter is transcribed into two single-stranded RNAs:
   - Smaller RNA is transported to the cytoplasm, where it is translated into viruscoded proteins.
   - Larger RNA is also transported to the same location in the cytoplasm, but it becomes encapsidated by coat protein subunits and is used as a template for reverse transcription into a complete virion dsDNA.

III. Replication of Single stranded DNA (ssDNA)

1. The method of replication of the single stranded DNA (ssDNA) of plant ssDNA viruses has not yet been determined with certainty.
2. There is some evidence, however, that the ssDNA replicates by forming a rolling circle that produces a multimeric (-) strand, which serves as a template for the production of multimeric (+) strands that are then cleaved to produce unit length (+) strands.
3. As soon as new viral nucleic acid is produced, some of it is translated, i.e., it induces the host cell to produce the protein molecules coded by its nucleic acid.

Virus protein synthesis

1. For virus protein synthesis, the part of the viral RNA coding for the viral protein plays the role of messenger RNA.
2. The virus utilizes the AMINO ACIDS, RIBOSOMES, and transfer RNAs of the HOST; however, it becomes its own blueprint (messenger RNA), and the proteins formed are for exclusive use by the virus as a coat or in other functions.
3. When new virus nucleic acid and virus protein subunits have been produced, the nucleic acid organizes the protein subunits around it, and the two are assembled together to form the complete virus particle, the virion.
4. The site(s) of the cell in which virus nucleic acid and protein are synthesized and in which these two components are assembled to produce the virions varies with the particular genus or family of the virus.
5. For most RNA viruses, the virus RNA, after it is freed from the protein coat, replicates itself in the CYTOPLASM, where it also serves as a messenger RNA and, in cooperation with the ribosomes and transfer RNAs, produces the virus protein subunits. The assembly of virions follows, also in the cytoplasm.
6. In other viruses, e.g., those with ssDNA, the synthesis of viral nucleic acid and protein, as well as their assembly into virions, seems to take

place in the NUCLEUS, from which the virus particles are then released into the cytoplasm.

7. The first intact virions appear in plant cells approximately 10 hours after inoculation.
8. The virus particles may exist singly or in groups and may form amorphous or crystalline inclusion bodies within the cell areas (cytoplasm, nucleus) in which they happen to be.

Translocation and Distribution of Viruses in Plants

1. When a virus infects a plant, it moves from one cell to another and multiplies in most, if not all, such cells.
2. Viruses move from cell to cell through the **PLASMODESMATA** connecting adjacent cells.
3. Viruses multiply in each parenchyma cell they infect.
4. In leaf parenchyma cells the virus moves approximately 1 millimeter, or 8 to 10 cells, per day.
5. In all economically important viral infections, viruses reach the phloem and through it are transported rapidly over long distances within the plant.
6. Most viruses, however, require 2 to 5 days or more to move out of an inoculated leaf.
7. Once the virus has entered the phloem, it moves rapidly in it toward growing regions (apical meristems) or other food-utilizing parts of the plant, such as tubers and rhizomes.
8. In the phloem, the virus spreads systemically throughout the plant and re enters the parenchyma cells adjacent to the phloem through plasmodesmata.
9. The development of local lesion symptoms is an indication of the localization of the virus within the lesion area.
10. In several diseases, however, the lesions continue to enlarge and, sometimes, the development of systemic symptoms follows, indicating that the virus continued to spread beyond the borders of the lesions.
11. In systemic virus infections, some viruses are limited to the phloem and to a few adjacent parenchyma cells.
12. Mosaic virus-infected plant cells have been estimated to contain between 100,000 and 10,000,000 virus particles per cell.
13. Systemic distribution of some viruses is quite thorough and may involve all living cells of a plant.

8

Transmission of Viruses
Virus - Vector Relationships

Transmission of Plant Viruses

- Plant virus rarely, if ever, come out of plant spontaneously.
- Plant viruses are not disseminated as such by wind or water.
- Viruses are transmitted from plant to plant in a number of ways.
- Vegetative propagation
- Mechanically through sap
- By seed
- By Pollen
- By Insect
- By Mites
- By Nematodes
- By Dodder
- By fungi

Transmission by Mechanical means &Vegetative Propagation

Mechanical Transmissions

- Such transmissions may takes place between closely spaced plants after a strong wind by contact.
- Plants are wounded during cultural operations.
- Virus infected sap adhering to the tools, worker hands or cloth accidently transmitted to the subsequently wounded plants.
- Example, Potato virus: (PV-X,TMV on Tobacco and Tomato)

Transmission by Vegetative Propagation

- Plants are propagated vegetatively by budding or grafting or by cutting or by the use of tubers, corms, bulbs or rhizome.

- Any virus present in the mother plant from which these organs are taken will almost always be transmitted to the progeny.
- Transmission of viruses may also occur through natural root grafts of adjacent plant.

Transmission by Seed, pollen and dodder

Transmission by Seed and pollen

- When viruses are transmitted by seeds, the seed is infected in the generative cells and the virus is maintained in the germ cells and sometimes, but less often, in the seed coat.
- When the growth and development of plants is delayed because of situations like unfavourable weather, there is an increase in the amount of virus infections in seeds.
- Seed transmission occurs because of a direct invasion of the embryo via the ovule or by an indirect route with an attack on the embryo mediated by infected gametes.
- Bean common mosaic virus is transmitted through seeds.

Dodder Transmission

- Several plant virus can be transmitted from one plant to another plant through the bridge formed between the two plants by twining the stem of parasitic plant, dodder (Cuscuta species).
- Examples: Green strain of cucumber mosaic virus on *N. glutinosa*, lucern (alfa-alfa), mosaic virus to tobacco, potato stem mottle virus to tobacco.
- Bennett (1940) showed dodder transmission in sugar beet curly top virus (BCTV) and cucumber mosaic virus.

Transmission by Insects

Specificity of Transmission

- Several insects transmit virus diseases.
- Virus transmission by the insects is highly specific.
- Transfer is not merely mechanical and the relationship between virus and vector is complex.
- Some easily transmissible viruses by mechanical means (by contact, on hands and equipments) are not transmitted by insects.
- Mechanically transmitted viruses like tobacco mosaic virus, potato virus X and white clover mosaic virus are not transmitted by insects.

- Viruses transmitted by leafhoppers are not transmitted by aphids.
- Strains of one virus (e.g., barely yellow dwarf virus) may have different insect species for their transmission.

Mouth Parts of Insects and Transmission

- Insects have different kinds of mouthparts and ways of feeding.
- **Biting and chewing mouthparts.**
    - Example; Beetles and Caterpillars .
    - They eat away the leaf pieces and damage the plants severely .
    - Cells which are touched by their mouthparts are destroyed.
    - Viruses coming through these mouthparts cannot establish on the leaves.
    - Hence transmission is rare among the biters and chewers.
    - Some viruses like turnip yellow mosaic virus and cowpea mosaic virus are transmitted by these beetles.
- **Piercing and sucking mouthparts.**
- Majority of the insects that transmit viruses have piercing and sucking mouthparts.
- Aphids and jassids have this kind of mouthparts.
- **Aphids**
    - Aphids make initial brief sap-sampling probes between epidermal cells of into them.
    - However, they take up food from phloem vessels only.
    - They need several minutes to hours to penetrate between plant cells rather than through cells to finally reach the phloem.
    - Aphids mostly transmit viruses from and into host parenchyma.
    - Acquisition of the virus and inoculation is in the initial probing period which is of a few seconds to some minutes.
- **Leafhoppers**
    - Mouthparts of leafhoppers are most robust compared to aphids.
    - They penetrate the plant cells and reach the phloem within a few minutes.
    - In the phloem, the sap flows under pressure and it will pass up into the insect's food canal.
    - Hence, the leafhoppers transmit the viruses found in the phloem of infected plants.

- **Thrips**
  - Have rasping and sucking mouthparts.
  - They feed by gasping open the epidermal cells or by rasping them and sucking up their contents.
  - Thus, thrips can transmit viruses found in epidermal cells also.

Relationship between Insect Vector and Virus

- Virus-vector relationship varies widely .
- It depend upon the duration of retention of the virus in the vector (persistence).
- In case of persistent viruses, the virus may simply circulate through the body of the vector or progagate also.

Classification of relationship between Insect Vector and Virus

1. Non-circulative Non-persistent
2. Non-circulative Semi-persistent
3. Circulative Non-propagative
4. Circulative Propagative Transmission

Non-circulative Non-persistent Transmission

- Virus is acquired and inoculated during brief feeding periods of a few seconds to some minutes, mostly during probes.
- Viruses are 'stylet-borne.
- Hence, transmission is non circulative non-persistent.
- Persistence in the vector is very brief.
- This method of transmission is only mechanical process of stylet contamination.
- Transmission is of low specificity.
- Aphids become infective immediately after virus uptake.
- Aphids such as *Myzus persicae, Aphis craccivora* and *Aphis gossypi* transmit several viruses.

Circulative Non-propagative (Persistent) Transmission

- Viruses taken up by the vector enter the alimentary canal, pass through the gut wall, circulate in body fluid (haemolymph) and contaminate the saliva.
- Insect becomes infective at the next feeding after a latent period.
- Infectivity is lost at moulting.

- Such uptake and inoculation happens only after long feeding times, usually about 15 minutes.
- Viruses are acquired from the phloem of vascular bundles.
- Persistent transmission is associated with highly specific-virus-vector relationships.
- Leafhoppers generally transmit viruses in persistent manner.

Circulative Propagative (persistent) Transmission

- Certain viruses are propagated inside the insect body.
- It may last till death.
- Such viruses can persist in populations without access to susceptible hosts.
- Once infective, the insects can transmit the virus to healthy plants throughout their life period.
- Even transovarial transfer of the viruses through insect eggs is found in leafhoppers.
- Example: Rice dwarf, Rice stripe, and Wheat striate mosaic viruses.

Non-circulative Semi-persistent

- Semi-persistent viruses can be transmitted mechanically also with some difficulty.
- Viruses are taken up from phloem by the insects along with food.
- They are adsorbed and gradually eluted from the line of pharynx.
- Viruses are transmitted through ejection or egestion at the starts of subsequent feeding.
- Example: Beet yellows virus

Insects which transmit Viruses

- Several insects belonging to different orders and families transmit virus.
- Orders transmit virus are given below:

  1. Order: Coleoptra

  2. Order: Homoptera

  3. Order: Thysanoptera

  4. Order: Orthoptera

  5. Order: Dermaptera

  6. Order: Heteroptera (true bugs)

  7. Order: Diptera (files)

Order: Homoptera

a) Family: Aphididae (Aphids)

Aphis craccivora - Groundnut rosette virus

Myzus persicae - Potato leaf roll virus

Toxoptera citricida - Citrus tristeza virus

Cavariella acgopodii - Carrot red leaf virus

Acyrthosiphon pisum - Bean leafroll virus

b) Family: Cicadellidae (Leafhoppers)

Nephotettix virescens - Rice tungro virus

Maize streak virus - Beet curly top

c) Family: Delphacidae (Delphacid planthoppers)

Rice hoja blanca virus - *Sogatodes oryzicola*

Sugaracane fiji disease - *Perkinsiella saccharida*

Maize rough dwarf virus - *Delphacodes propingua*

d) Family: Aleurodidae (Whiteflies)

Bemisia tabaci - Bhendi vein clearing virus, Tobacco leaf curl virus, Tapioca mosaic virus

Trialeurodes vaporariorum - Beet pseudo-yellows virus

e) Family: Membracidae (Treehoppers)

Micrutalis malleifera - Pseudo curly top virus

f) Family: Pseudococcidae (Mealy bugs)

Planococcus spp. - Cacao swollen shoot

Order: Coleoptra

a) Family: Chrysomelidae (Leaf beetles)

Ceratonia spp., Diabrotica spp., *Ootheca mutabilis* - Cowpea mosaic virus

Podagria spp. - Bbendi mosaic virus
Psylliodes spp. - Turnip yellow mosaic virus

b) Family: Coccinellidae (Lady beetles)

Epilachna spp. - Cowpea mosaic virus

c) Family: Curculionidae (Snout beetles or Weevils)

Apion spp., *Sitona lineata* - Broadbean stain virus

d) Family: Meloidae (Blister beetles)

Epicauta vittata: Bean pod mottle virus

Order: Thysanoptera

Family: Thripidae (Thrips)

Frankliniella spp.-Tomato spotted wilt

Thrips tabaci -Tomato spotted wilt

Order: Orthoptera

Family: Acrididae (Grasshoppers)

Melanoplus differentialis: Potato virus X, Tobacco mosaic virus (Mechanical transmission)

Order: Dermaptera

Family: Forficulidae (Earwigs)

Foficula auricularia-Turnip yellow mosaic

Order: Heteroptera (True bugs)

a) Family: Miridae (Plant bugs)

Lygus protensis- Spinach blight virus

b) Family: Tingidae (Lace bugs)

Piesma quardratum- Beet leaf curl virus

c) Family: Lygaeidae (Lygaeid bugs); Nysius spp.- Centrosema mosaic virus

Order: Diptera (Files)

Family: Agromyzidac (Leaf miner flies)

Liriomyza langei: Sowbane mosaic virus (Mechanical transmission)

Transmission by Mites

- Mites have puncturing and sucking mouthparts.
- Mites belonging to two families are known to transmit viruses.
- **Eriophyidae**.
- Eryophid mites belong to the family.
- They are invisible to the naked eye and they often hide between hairs of leaves and in buds.
- They have slender stylets with which they puncture epidermal cells.

- The two pads at the tip of the rostrum belp in bringing saliva to the stylets and in sucking up plant sap.
- They don't have much movement and are carried by wind.
- *Abacarus hystrix* transmits ryegrass mosaic virus.
- *Aceria cajani* transmits pigeon-pea sterility mosaic virus.
- **Tetranychoidae**.
- Spider mites belong to the family Tetranychoidae.
- *Brevipalpus obovanus* transmits Citrus leprosis virus while *B. phoenicis* transmits coffee ringspot virus.

Transmission by Nematodes

- Hewitt and colleagues (1958)first showed that Fan leaf virus of grapevines is transmitted by dagger nematode, *Xiphinema index*.
- Most nematodes parasitic on green plants belong to order Tylenchyda, but non of this group has yet been found to be a vector virus.
- All species of nematodes known to transmit viruses are members of the order Dorylaimida.
- It belong to the five genera: *Xiphineama, Longidorus, Paralongidorus, Trichodorus* and *Paratrichodorus.*
- Longidorus and Xiphinema species transmit NEPO viruses such as the Grapevine fan leaf virus.
- Trichodorus and Paratrichodorus species transmit Tobraviruses like Tobacco rattle virus and Pea early browning virus.

Examples of Viruses Transmitted by Nematodes

Sr. No	Group	Nematode Virus Vectors	Example
1	NEPO: Nematode transmitted polyhedral viruses	*Xiphinema index*	• Grape vine fan leaf virus
		Xiphinema coxy	• Cherry leaf roll virus
		Xiphinema americaum	• Ring spot of tobacco virus • Tomato, soybean bud blight virus • Bringal mosaic virus
2	NETU: Nematode transmitted tubular viruses	*Longidours*	• Rasberry ring spot virus, Tomato black ring virus
		Trichodours and Para trichodorous	• Rod shaped viruses Tobacco rattle virus • Pea early browning virus

Transmission by Fungi

- Some soil-borne fungal pathogens transmit plant viruses.
- Olpidium spp. transmit cucumber mosaic virus, tobacco necrosis virus and redclover necrotic mosaic virus.
- *Polymyxa graminis* transmits wheat mosaic virus while *Polymyxa betae* (same family) transmits beet necrotic yellow vein virus.
- *Spongospora subterranea* transmit beet necrotic yellow vein virus.
- All these fungi are pathogens of the host which carry the viruses.
- Zoospores of the fungi are released from the host and the zoospores carry the virus and transmit it to the susceptible hosts during their infection process.
- Plant viruses are also carried on the outside of the fungi in few cases.
- Examples are tobacco necrosis virus and cucumber mosaic virus.
- The viruses like lettuce big vein vinus are found inside the zoospores.
- They persist for years in viable resting sporangia.
- The two types of transmission by fungi can be considered as non-persistent and persistent transmission.

Examples of Viruses Transmitted by Fungi

Sr.No	Fungus Virus Vectors	Example
1	*Olpidium brassicae*	• Lettuce big vein virus • Tobacco necrosis virus • Cucumeber necrosis virus • Tobacco stunt virus
2	*Synchytrium endobioticum*	• PU-X virus
3	*Polymyxa graminis*	• Wheat mosaic virus • Beat necrotic yellow vein virus
4	*Spongospora subteranea*	• Potato mop top virus

Horizontal and Vertical Transmission of Virus

Plant viruses spread from plant to plant by two major routes

1. Horizontal transmission
2. Vertical transmission

1. Horizontal transmission

- Route of viral transmission in which an organism receives the virus from an external source.
- Plants are more susceptible to viral infection if their protective epidermal layer is damaged.

- Insects may be *vectors* that transmit viruses from plant to plant and can inject the virus directly into the cytoplasm.
- By using contaminated tools, gardeners and farmers may transmit plant viruses.

2. Vertical transmission

- Route of viral transmission in which an organism inherits a viral infection from its parent.
- Can occur in asexual propagation of infected plants (e.g., by taking cuttings).
- Can occur in sexual reproduction via infected seeds.

9

Mycoviruses and Baculoviruses

Mycoviruses

- Study of mycoviruses are called "mycovirology".
- Viruses that infects fungi are called mycovirus.
- Known as mycophages.
- First reported by Hollings.
- Majority of mycoviruses genome is made up of dsRNA (roughly 70%).
- Few had (+) ss RNA (30 %).
- No mycovirus reported to have dsDNA.
- Transmitted form one fungus to other, without being released as a free particle into external environment.
- Economic impact of mycoviruses on fungi was first recorded in cultivated mushrooms (*Agaricus bisporus*) and called the La France disease.
- La France disease also known as X disease or watery stripe or dieback or brown disease.
- Fungal viruses are variously implicated in reducing the yields of cultivated mushroom and pathogenicity of parasitic fungi.
- Most of the fungal viruses are icosahedral and ranges from 25-45 nm in diameter.
- Mycoviruses are common in fungi and are found in all four phyla of the true fungi: Chytridiomycota, Zygomycota, Ascomycota and Basidiomycota

Characters of Mycoviruses

- Most mycoviruses lack an extracellular route for infection.
- Transmitted intercellularly only through cell division, sporulation, and cell fusion.
- Apparently lack a movement protein, which is essential for the life cycle of animal and plant viruses.

Baculoviruses

- Baculoviridae is a family of viruses that infects insects.
- Arthropods, lepidoptera, hymenoptera, and diptera serve as natural hosts.
- Genome is composed of double-stranded DNA that codes for genes needed for virus establishment and reproduction.
- Majority of baculoviruses used as biological control agents are in the genus Nucleopolyhedrovirus.
- They have no negative impacts on plants, mammals, birds, fish, or even on non-target insects.
- This is especially desirable when beneficial insects are being conserved to aid in an overall IPM program, or when an ecologically sensitive area is treated.
- USDA Forest Service currently uses the gypsy moth nuclear polyhedrosis virus (LdNPV) to aerially spray thousands of acres of forest each year.
- This product, registered as Gypchek, is effective against gypsy moths but leaves all other animals unharmed

Uses of Baculoviruses as Biological Insecticides

Commodity	Insect pest	Virus used	Virus product
Coton and vegetables	Tobacco budworm Helicoverpa tea, and Cotton bollworm *Heliothis vinescens*	*Helicoverpa zea* nuclear polyhedrosis virus	Gemstar LC. Biotrol, Elcar
Vegetable crops, greenhouse	Beet armyworm (*Spodoptera exigua)*	*Spodoptera exigua* nuclear polyhedrosis virus	Spod-X
Forest Habitat, Lumbe	Douglas fir tussock moth (*Orgyia pseudotsugata*)	Orgyiapseudotsugata nuclear polyhedrosis vinis	TM Biocontrol
Apple, pear, walnut and plum	Codling moth	Codling moth granulosis virus	Cyd-Xe
Cabbage, tomatoes, cotton, (and see pests in nest column)	Cabbage moth. American bollwom diamondback moth, Potato tuber moth, and Grape berry moth	Cabbage army worm nuclear polyhedrosis virus	Mamestrin
Cotton, Tomatoes	*Spodoptera littoralis*	*Spodoptera littoralis* nuclear polyhedrusis virus	Spodopterin

10

VIROIDS

- Viroids are small, low molecular weight ribonucleic acids that can infect plant cells, replicate themselves, and cause disease.
- Viroids differ from viruses in at least two main characteristics:
- The size of RNA in viroids, which consists of 250 to 370 bases, is much smaller compared to that in viruses, which is 4 to 20 kilobases.
- Virus RNA is enclosed in a protein coat whereas viroids lack protein coat and apparently exist as free (naked) RNA.
- Due to their small size (250–370 nucleotides), viroids lack sufficient information to code for even one protein, even for a replicase enzyme required to replicate the viroid.
- Existence of viroids as free RNAs rather than as nucleoproteins necessitates the use of phenol in the sap to inactivate the plant ribonucleases and makes their visualization with an electron microscope.

Characteristics of Viroids

1. Viroids are circular, single-stranded RNA molecules with extensive base pairing in parts of the RNA strand.
2. The base pairing results in some sort of hairpin structure with single-stranded and double-stranded regions of the same viroid and contributes to the stability of the RNA, given that it lacks a protein coat.
3. It appears that, in its double-stranded form, each viroid consists of five structural regions: a left and a right terminal region, a pathogenicity region, a conserved central region, and a variable region.
4. Terminal and pathogenicity regions determine the pathogenicity of a viroid, i.e., its ability to infect and multiply, and also the severity of the symptoms that will develop on the host plants.
5. Severity of the symptoms, however, can be altered by changes in one or two bases in these regions.
6. Other two regions of viroids, the conserved central region and the variable region, have not been implicated in any function of viroids.

7. Viroids seem to be associated with cell nuclei, particularly the chromatin, and possibly with the endomembrane system of the cell.
8. How viroids replicate themselves is still not known.
9. Their small size is sufficient to code for a very small protein, but such a protein would be considerably smaller than known RNA polymerase (replicase) subunits and would therefore be unable to carry out replication of the viroid.
10. In addition, viroids have been shown to be inactive as a messenger RNA in all *in vitro* protein-synthesizing systems tested.
11. Also, no new proteins could be detected in viroid-infected plants.

Viroid Replication

1. Viroids replicate by direct RNA copying in which all components required for viroid replication, including the RNA polymerase, are provided by the host.
2. During viroid replication, the circular (+) strand of the viroid is replicated while it acts as a rolling drum producing multimeric linear strands of (-) RNA .
3. Linear (-) strand then serves as a template for replication of multimeric strands of (+) RNA.
4. The (+) RNA is subsequently processed (cleaved) by enzymes that release linear, unit-length viroid (+) RNAs, which circularize and produce many copies of the original viroid RNA .

Taxonomy (Grouping) of Viroids

1. Taxonomy of viroids is based on the absence in some of them (the avocado sunblotch viroid group, group A, or Avsunviroids) of a conserved central region or the presence in them (the potato spindle tuber viroid group, group B, or Pospiviroids) of a central conserved region.
2. Avocado sunblotch viroid (ASBVd) group has only four members, whereas the potato spindle tuber viroid (PSTVd) group has all the rest of the 40 viroids.
3. All Avsun viroids have a ribozyme activity that enables them to self-cleave their RNA multimers during viroid replication.
4. Avsun viroids replicate in chloroplasts, whereas Pospiviroids replicate in the nucleus and nucleolus.
5. Both groups are subdivided into subgroups depending on sequence similarities in the conserved central region.

Classification

Viroids

ASBVd group or Avsunviroids

Avsunviroideae

Avsunviroid

Avocado sunblotch viroid

Pelamoviroid

Chrysanthemum chlorotic mottle viroid

Peach latent mosaic viroid

PSTVd group or Pospiviroids

Pospiviroideae

Pospiviroid subgroup

Potato spindle tuber viroid

Chrysanthemum stunt viroid

Citrus exocortis viroid

Columnea latent viroid

Iresine viroid 1

Mexican papita viroid

Tomato apical stunt viroid

Tomato planta macho viroid

Apscaviroid subgroup

Apple scar skin viroid

Apple dimple fruit viroid

Australian grapevine viroid

Citrus bent leaf viroid

Citrus viroid III

Grapevine yellow speckle viroid

Grapevine yellow speckle viroid 1

Grapevine yellow speckle viroid 2

Pear blister canker viroid

Cocadviroid subgroup

Coconut cadang-cadang viroid

Citrus viroid 4

Coconut tinangaja viroid

Hop latent viroid

Coleviroid subgroup

Coleus blumei viroid

C. blumei viroid 1

C. blumei viroid 2

C. blumei viroid 3

Hostuviroid subgroup

Hop stunt viroid

Unassigned viroids

Apple fruit crinkle viroid

Cherry small circular viroid-like RNA

Citrus viroid Ia

Citrus viroid II

Citrus viroid OS

Citrus viroid-I-LSS

Coleus yellow viroid

Tomato chlorotic dwarf viroid

How viroids cause disease

1. How viroids cause disease is also not known.
2. Viroid diseases show a variety of symptoms that resemble those caused by virus infections.
3. The amount of viroids formed in cells seems to be extremely small, and it is therefore unlikely that they cause a shortage of RNA nucleotides in cells.
4. Even one or two base changes at specific sites of the viroid are sufficient to change the disease from mild to severe and vice versa.
5. Thus, viroids apparently interfere with the host metabolism in ways resembling those of viruses, but which ways are also unclear.
6. It has been shown that bothvirus-specific RNAs synthesized during infection and viroid RNA *in vitro activate a protein kinase enzyme,* which in turn activates other cellular enzymes while it impedes the initiation of protein synthesis.
7. As viroid strains that cause mild to severe plant symptoms activate the protein kinase more than 10 times as much as mild strainsit is possible that activation of the protein kinase represents the triggering event in viroid pathogenesis and in disease development by the plant.

Transmission of Viroids

1. Viroids are spread from diseased to healthy plants primarily by mechanical means, i.e., through sap carried on hands or tools during propagation or cultural practices and, of course, by vegetative propagation.
2. Some, such as potato spindle tuber, chrysanthemum stunt, and chrysanthemum chlorotic mottle viroids, are transmitted through sap quite readily, whereas others, such as citrus excortis viroid, are transmitted through sap with some difficulty.
3. Several viroids, e.g., those causing potato spindle tuber, cadang-cadang, tomato bunchy top, and apple scar skin, appear to be transmitted through the pollen and seed, but the rates of such transmission are usually very small.
4. No specific insect or other vectors of viroids are known, although viroids seem to be transmitted on the mouthparts or feet of some insects.

Survival of Viroids

1. Viroids apparently survive in nature outside the host or in dead plant matter for periods of time varying from a few minutes to a few months.
2. Generally, they seem to overwinter and oversummer in perennial hosts, which include the main hosts of almost all known viroids.

3. Viroids are usually quite resistant to high temperatures and cannot be inactivated in infected plants by heat treatment.

Management

1. Use of viroid-free propagating stock,
2. Removal and destruction of viroid-infected plants.
3. Washing of hands or sterilizing of tools after handling viroid infected plants before moving on to healthy plants.

Unit 7: Plant Disease Epidemiology

1

Concepts in Epidemiology

- Vanderplank called epidemiology the *science of disease in population.*
- Vanderplank puts emphasis on *population phenomena.*
- Concerned with *host-pathogen interactions* that lead to disease and crop losses.
- These interactions are often environmentally dependent.
- So epidemiology deals with effects of the biotic and abiotic environments.
- It also deals with the effects of man on disease and sometimes with the effects of plant disease on mankind.
- All of this can be symbolized by the *disease tetrahedron.*

History (Ancient to Modern times)

- **Hippocrates** (~400 BC): First use of "epidemic", widespread disease (human diseases).
- **Theophrastus** (~340 BC): Plant diseases in fields, Environmental influences.
- **Pliny** (~50 AD): Plant diseases; soil; climate.
- **Duhamel de Monceau** (1728 AD): Disease progress curves, Comparison of plant and animal epidemics.
- **Late 19th Century and forward...**
- **Kuhn (1858) -** Wrote 1st textbook of plant pathology
- **Ward (1901):** Wrote book "Diseases in Plants" emphasized ecology (populations) of disease
- **Jones (1913)** - role of the environment
- **Gaumann (1946):** "Principles of Plant Infection" -Disease spread, -Conditions leading to an epidemic, -*'Infection Chain' (= disease cycle),* -compare with medicine (diseases of humans).

Purpose of Study of Epidemiology

The study of epidemiology serves at least two purposes;

- Understanding of the behaviour of disease in time and space
- Using that understanding to manage disease.

Elements of an Epidemic

1. Host
2. Pathogen
3. Environment; Interactions of the 3 main components are described by the **disease triangle**. Disease development is also affected by
4. Time
5. Humans

Disease Tetrahedron

Interactions of the 5components are described by the disease pyramid.

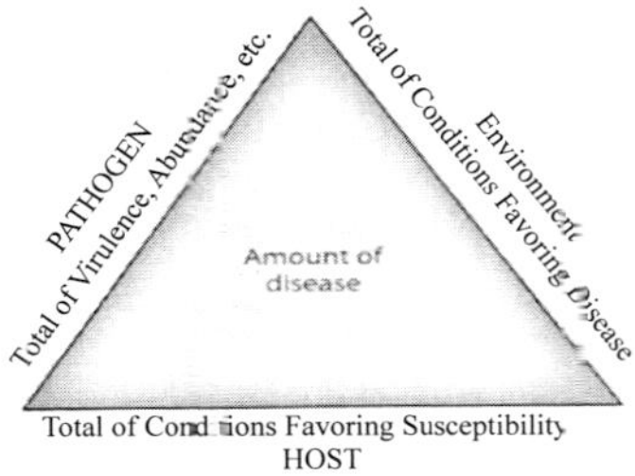

Disease Triangle

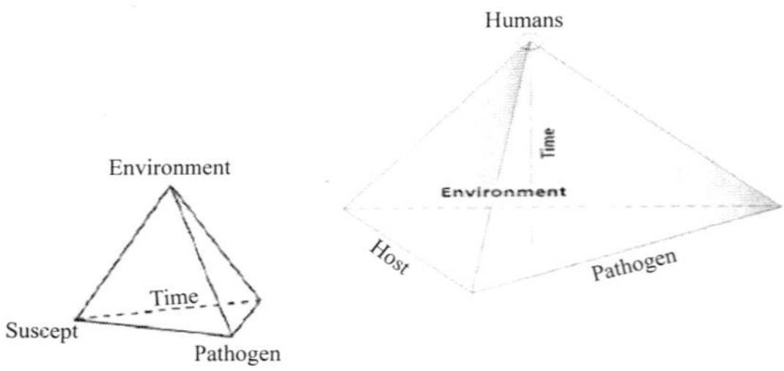

Disease Tetrahedron

Disease progress curve for a typical monocyclic pathogen

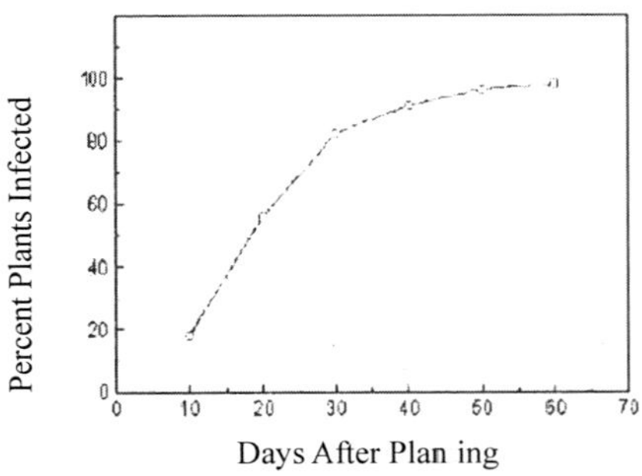

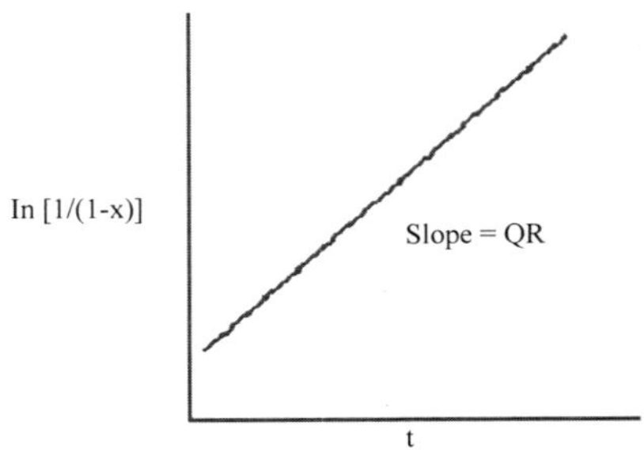

Progress curve for polycyclic epidemic

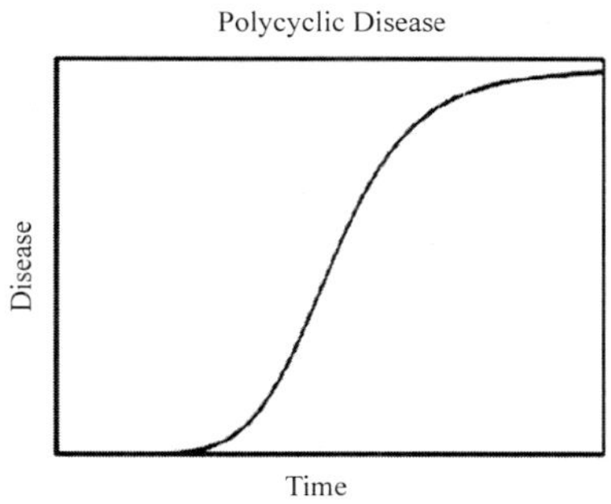

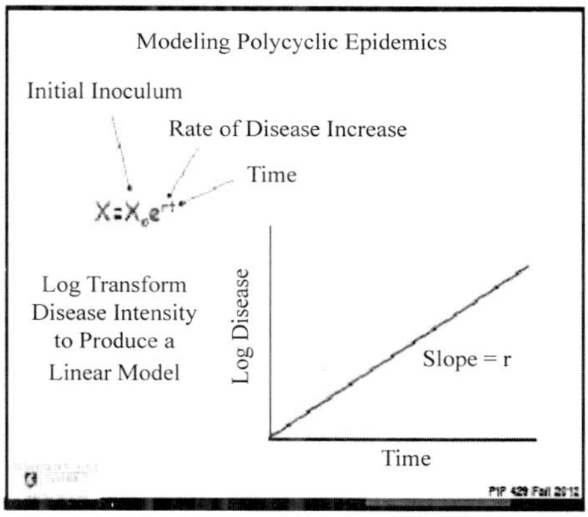

Comparison of Epidemic Models

Parameter	Monocyclic Model	Polycyclic Model
Reproduction	Low reproduction and death rate	High reproduction and death rate
Inoculum dispersal	Soil	Wind borne, Vectors
Examples	Root diseases, Wilt diseases, Stem diseases	Leaf spot, Leaf blight, Leaf rust, Powdery mildew
Management	Use resistant varieties, Stop pathogen reproduction, Spray fungicides	Reduce initial inoculums Destroy diseased plant or debris
Epidemic rate (r)	Eg. Potato late blight R= 0.3 -0.5/day	Ex. Verticillium wilt R= 0.02/day

Measuring Disease in a Population

Disease incidence

1. Actual number or proportion of plants diseased.
2. Number diseased out of total number of plants observed.

Disease severity

Area of plant tissue affected by disease

1. For many diseases, severity is the area of plant surface covered by lesions
2. Measured using assessment scales or by determining the area under a disease progress curve (AUDPC)

Yield loss

1. The proportion of yield that the grower will not be able to harvest due to disease.
2. Results in economic loss.

Disease gradients

1. Change in disease severity along a straight line away from the source of inoculum.

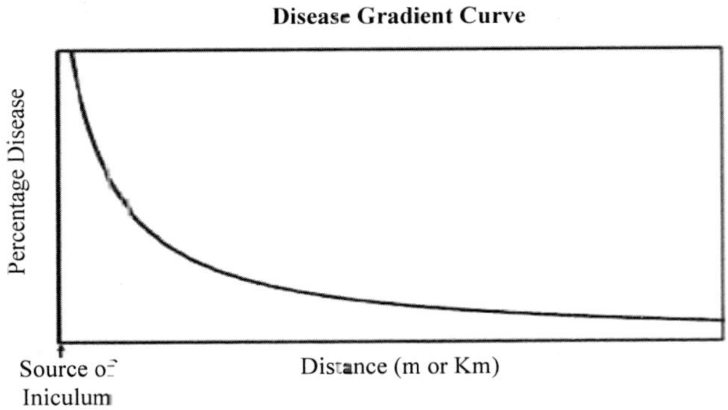

Epidemic Management

A. Ways to reduce disease(inoculum) at beginning (xo)

1. Affects monocyclic and polycyclic diseases
2. Sanitation
3. Certified seed
4. Seed treatments
5. Quarantine
6. Host plant resistance
7. Fumigation

B. Ways to decrease the rate of disease development (infection rate) (r)

1. Change the environment
2. Fertilizer application
3. Pesticide Application
4. Host plant resistance
5. Ways to change t
6. Harvest early before disease becomes severe.
7. Plant early (cereal cyst nematode)

2

Factors Affecting Disease Development

Pathogen Factors	Host Factors	Environmental Factors
Presence of pathogen	Type of crop	Temperature
Mode of dispersal of pathogen	Degree of genetic uniformity of host plants	Rainfall / Dew
High birth rate of pathogen	Genetic make-up	Leaf wetness period
Low death rate of pathogen	Host age	Soil properties
Survival efficiency	Host nutrition	Wind
Reproductive fitness	Host population structure	Fire history
Adaptability of pathogen	Nature of propagation	

Host Factors

a. Type of crop

- Epidemics generally develop much more rapidly in annual crops (cereal. pulses, oilseeds and vegetable crops) than perennial crops (fruit and forest trees).
- Some epidemics of fruit and forest trees, for instance, pear decline, tristeza in citrus, Dutch elm disease and chestnut blight, takes year to develop.

b. Degree of genetic uniformity of host plants

- A severe plant disease epidemic may develop when genetically uniform plants are grown over large area.
- Lowest rates of epidemic development generally occur in cross pollinated crops, intermediate rates in self-pollinated crops and the highest rates in vegetative propagated crops.
- Greater likely hood exists that a new pathogen race will appear that can attack their genome and result in an epidemic.
- For instance, in southern corn leaf blight on corn carrying Texas male sterile cytoplasm.

c. Level of genetic resistance or susceptibility of the host

- Vertical and horizontal resistance have distinct bearing on development and progress of epidemic.

- Hosts plants carrying horizontal resistance will probably become infected, but the rate at which the disease and the epidemic will develop depends on the level of resistance and the environmental conditions.
- Host plants carrying vertical resistance do not allow a pathogen to become established in them, and thus no epidemic can develop.
- Susceptible host plants lacking genes for resistance against the pathogen provide the ideal substrate for establishment and development of new infections.

d. Host age

- Plant change in their susceptibility to disease with age.
- For instance, damping off and roots rots, downy mildews, peach leaf curl, rusts, smuts, bacterial blights and viral infections, the host plants are susceptible only during their growth period and becomes resistant during adult stage.
- In some other diseases such as flower or fruit blights caused by fungal pathogens like *Alternaria, Penicillium, Glomerella* and *Monilinia* and in all post-harvest infections, plant parts(fruits) are resistant during growth and early adult stage but become susceptible near riepning.

e. Host nutrition

- Balanced nutrition is required for proper growth and development of plants.
- Under good nutritional conditions, biotrophic fungi have a relative advantage.
- Under poor nutritional conditions necrotrophic and perthotrophic fungi have a relative advantage.
- The type and amount of fertilizers have influence on the development of plant diseases.

f. Host population structure

- Crop rotation and sequence and crop density have major bearing on the type and amount of plant diseases.
- High crop density will influence microclimate and will make infection and dispersal of inoculums more effective.
- Both these factors will help the development of disease.

g. Nature of propagation

- Nature of propagation affects the development of disease.
- Chances of disease are more if plants are vegetatively propagated.

Pathogen Factors

a. Level of virulence

- Virulent pathogens capable of rapidly infecting the host.
- Ensure faster production of larger amounts of inoculums.
- Thereby, disease, then pathogens of lesser virulence.

b. Mode of dispersal of pathogen

- Rapid and easy dispersal of the pathogen inoculums from its source to healthy hosts in the field may help frequent and wide spread epidemic.
- Epidemics of diseases like mildew, rusts and leaf spots occur usually as the spores of their pathogen are released into the air and get disseminated quickly by it over distances up to several miles.
- Contrary to it, soil borne pathogens like *Fusarium, Verticilium*, etc., and most nematodes usually have 2 to 4 reproductive cycles per growing season.
- Dispersal of such pathogens is limited both in space and time, only localized and slower developing epidemics are caused.

c. High birth rate of pathogen

- High birth rate or fast reproductive cycle of pathogen is key factor for establishment and development of epidemics.
- Some fungal pathogen have capacity to produce enormous quantity of spores asexually.
- These spores, if dispersed fast over long distances and favored by conducive weather conditions, can result in epidemics in their respective areas.

d. Low death rate of pathogen

- Epidemics attributed to low death rate of pathogens are those in which the causal organism is systemic and protected by the plant tissues.
- Thus, chances of high mortality are considerably reduced and ultimately greater chance of widespread infections in a given area exist.

e. Adaptability of pathogen

- For pathogens having capacity to acclimatize to adverse conditions, the occurrence of epidemics is about certain.
- The units of propagules produced by the pathogen are dispersed by external agencies which must be available if epidemics are to develop.

Role of Environmental factors

- Presence of a pathogen against a particular plant will generally not cause serious disease unless the environmental conditions are favorable.
- This includes
    - Aerial environment
    - Soil (*edaphic*) environment.

Aerial Environment

a) Temperature

- Temperature affects the
    - Incubation (Period of time between infection with pathogens and the appearance of symptom of disease).
    - Latent period (the time between infection and the appearance of disease symptoms), the generation time (the time between infection and sporulation).
    - Infectious period (the time during which the pathogen keeps producing propagules).
- The disease cycle speeds up at higher temperatures, resulting in faster development of epidemics.
- Period of leaf wetness, combined with temperature information can be used to predict outbreaks of some diseases (infection periods) and be used to time preventative treatments, such as spraying.

b) Moisture

- Moisture is mainly important to pathogenic fungi and bacteria.
- Rain splash plays an significant role in the dispersal of some fungi and nearly all bacteria.
- Because the process of germination and infection takes time, the duration of leaf wetness also influences the success of the infection.
- Generally, a longer period of leaf wetness is desired to establish an infection in cooler temperatures, as germination and infection are normally accelerated in warmer conditions.
- The duration necessary for infection varies with temperature.

c). Pollution

- A lately recognized aspect of the aerial environment that can influence disease in plants is air **pollution.**
- A high concentration of pollutants can affect disease development and, in extreme cases, damage the plants directly by causing acid rain.

Soil Environment

- Soil environment affects soil-borne diseases, largely by determining the amount of moisture available to pathogens for germination, survival and motility.
- Germination and infection success also rely on the temperature of the soil.
- The fertility and organic matter content of the soil can affect the development of disease.
- Plant defenses are weakened by nutrient deficiency, although some pathogens, such as rusts and powdery mildews, thrive on well-nourished plants.
- Other diseases thrive in soils that are specifically low in organic matter.

3

Types & Course of Epidemics

Types of Epidemics

Monocyclic Epidemics

- Caused by pathogens with a low birth rate and death rate.
- They only have one infection cycle per season.
- They are typical of soil born diseases such as *Fusarium* wilt of pigeonpea.

Polycyclic Epidemics

- Caused by pathogens capable of several infection cycles a season.
- These are most often caused by airborne diseases such as powdery mildew.

Polyetic Epidemics

- Some diseases takes several years from the time of infection until symptom develop and pathogen reproduction.
- Can be caused by both monocylcic and polycyclic pathogens.
- **Example**
    - Apple powdery mildew is an example of a polyetic epidemic caused by a polycyclic pathogen.
    - Dutch Elm disease a polyetic epidemic caused by a monocyclic pathogen.

Differences between monocyclic and polycyclic plant disease epidemics

Characters	Monocyclic	Polycyclic
Birth rate	Low	High
Death rate	Low	High
Rate of increases	Like simple interest	Like compound interest
Survival	Long lived	Short lived
Reproduction	Reproduces only one generation per growing season	Reproduce new batches of propagules several times per growing season

Terminology Used in Epidemiology

- **Disease progress curve**
  - Periodic observations on disease severity, in any field or area, when plotted against time on a graph paper gives curved lines representing progress of disease over time.
  - These cumulative disease curves are known as Disease progress curve (DPC).
- **A saturation curve**-monocyclic disease
- **A sigmoid curve**- polycyclic disease
- **A bimodal curve**-polycyclic disease affecting different organs
- **Disease-gradient curve;** A graph that plots disease vs. distance from an inoculum source.
- **Epidemic;** A disease increase in a population; usually a widespread and severe outbreak of disease.
- **Epidemiology;** The study of factors affecting the outbreak and spread of disease.

Course of epidemic

Course of epidemic follows two distinct phases viz.

1. Progressively destructive phase
2. Decline phase

i. Progressively destructive phase

- During early stage, an epidemic spreads briskly causing diseases in new hosts.
- Some epidemics develop rapidly while others develop slowly.
- Slow epidemics (or epiphytotics) generally occur among population caused by systemic pathogens.
- Rapid epiphytotics are greatly caused by air borne pathgogens.
- Rapid epiphytotics are greatly influenced by environmental factors.

ii. Decline phase

- After development of a saturation stage it shows a decline by itself.
- No epidemics may be due to non-availability of susceptible stages of the crop, unfavourable weather conditions and reduction in aggressiveness of the pathogens.
- Normally the hosts are prone to the disease at a specific developing stage.

- Once this stage is crossed in a plant it's proneness to infections is reduced or completely lost. Under the conditions the epidemic declines.
- The decline in the epidemic may also be due to unfavourable weather conditions for disease development.
- As a result future spread of the disease will be checked and the epidemic will decline.

Slow & Rapid Epiphytotics

- Epidemic may develop slowly and is called 'tardive'.
- Epidemic which develops rapidly is called 'explosive' epidemic may occur.
- The form of epidemic is decided by the nature of the pathogen, host and the weather.

i. Rapid epiphytotics

- Occur among annual crops.
- Caused by non-systemic pathogens with high birth rate.
- Several generations of the pathogen is produced within a short time.
- Largely governed by environmental factors compared to slow epiphytotics.
- Disease increase is rapid and the disease rises to a beak in short time and then show sharp decline.
- The decline is due to unaffordable weather or when the host becomes resistant due to maturity or due to restricted dispersal of propagules of pathogen. e.g.,apple scab.
- This type of epiphytotic is controlled by protective spraying or dusting with chemicals.

ii. Slow epiphytotics

- Occur among perennial (tree) populations.
- Infected host survives for several years before dying.
- Most of the characters of a simple interest disease are found in slow epiphytotics. The causal agent is mostly systemic.
- Pathogen multiplies slowly.
- Their movement from plant to plant is much slower.
- They are low death rate pathogen.
- In slow epiphytotics, crop sanitation is the best method. e.g., Swollen shoot of cocoa.

4

Modeling in Plant Pathology

- Models are simple representation of otherwise more complex real systems.
- The model provide opportunity for better understanding and system management.
- The selection of suitable model is determined by the purpose and goals which must be defined very precisely.
- Modeling in plant pathology started with pioneering synthesis of Vander plank (1963) and aerobiology of Gregory (1961).
- Kranz (1974) stated that a model might be verbal statement, a hypothesis, a theory or a law.

Types of Models

1. Conceptual Models
2. Analytical Models
3. Predictive Models
4. Simulation Models

1. Conceptual Models

- These models were conceived during early stages of introduction of models in plant pathology.
- They were fundamentally qualitative, without much quantification.
- The relationships and interactions were displayed in diagrammatic form of flowchart manner.
- Example, Disease pyramid, Components of pathogenesis, gene for gene hypothesis.

2. Analytical Models

- Vanderplank (1963) used differential equations for calculation of pathogen population growth.
- Equation of single and multiple cycle diseases, threshold theorem and fitting curves to empirical data are common analytical models used in plant pathology.

3. Predictive Models

- Those in which the emphasis is placed on those variables which have the greatest value in forecasting the likely course of events.
- These are generally used in estimation of yields and forecasting diseases.
- Regression and differential equations are generally employed.
- They are considered a boon to growers for bringing precision in management decisions.
- Examples,
- BLITECAST in USA and PHYTOPROG in West Germany in giving shot term warnings of potato late blight attacks.
- FAST is the computer model for forecasting early blight in tomato.

4. Simulation Models

- Simulation models try to simulate or mimic the real life situations under study.
- When a computer is provided data describing the diverse subcomponents of the epidemic and measures of control at specific points in time.
- It then provides back the continuous information regarding not only the spread and severity of the disease over time, but also the final crop and economic losses likely to occur due to disease under the conditions of the epidemic as provided to the computer.
- The first computer simulation programme, called EPIDEM, was written in 1969 to simulate early blight epidemics of tomato and potato.
- **Examples** of other simulators are:

Name of the programme	Disease	Pathogen
EPIPRE	Cereal rusts and aphids	*Puccinica graminis tritici*
EPICORN	Southern corn leaf blight of maize	*Helminthosporium maydis*
CERCOS	Leaf blight of celery	*Cercospora spp.*
MYCOS	Blight of chrysanthemum	*Mycosphaerella spp.*
EPIVEN	Apple scab	*Venturia inaequalis*
EPIDEMIC	Stripe rust	*Puccinia striiformis*
TOM-CAST	Early blight of tomato	*Alternaria solani*
BLIGHT-CAST	Late blight of potato	*Phytophthora infestans*
PLASMO	Downey mildew of grapes	*Plasmopara viticola*
SIMCAST	Late blight of potato	*Phytophthora infestans*
NEGFRY	Late blight of potato	*Phytophthora infestans*
EPIVET	Viral diseases of potato	Contact and aphid transmitted viruses

5

Plant Disease Forecasting

Definition

Forecasting involves all the activities in ascertaining and notifying the grower of community that conditions are sufficiently favourable for certain diseases, that application of control measures will result in economic gain or on the other hand, and just as important that the amount expected is unlikely to be enough to justify the expenditure of time , energy and money for control.

Positive forecast: Employs need based chemical sprays, provides adequate protection to crop and reduces damage to environment.

Negative forecast: Avoids unnecessary chemical sprays, no risk to the crop health and no disruption of environment.

Importance of Forecasting

- Forecasting informs the growers that weather conditions are not favorable and the disease is unlikely to be intense enough. This helps growers to save the expenditure in terms of energy, time and money by not applying unnecessary management measures.
- It informs the growers about when the conditions are going to be sufficiently favorable for economically important diseases of an agroclimatic zone. This help growers to apply disease management timely that results in economic gain.
- It helps growers to plan advance preventive measures against likely losses due to the occurrence of economically important severe diseases.
- Government and other organizations initiate necessary steps for the timely stocking of chemicals, equipments, etc. when they are timely warned by the possibility of an epidemic by the forecasting.

Requirements for Disease Forecasting

- Disease must cause economically significant damage in terms of yield loss or quality.
- Control measures must be available at an economically acceptable cost.
- Disease should not be regular feature.

- Reliable means of communication with farmers.
- Growers must have sufficient manpower and equipments to apply control measures when disease warning is given.
- Farmers should be adaptive and have purchase power.

Information's Needed for Disease Forecasting

1. Host Factors

a. Prevalence of susceptible varieties in the given locality.
b. Response of host at different stages of the growth to the activity of pathogen.
c. Density and distribution of the host in a given locality.

2. Pathogen factors

a. Amount of primary inoculum in the air, soil or planting material.
b. Dispersal of inoculums.
c. Spore germination.
d. Infection.
e. Incubation period.
f. Sporulation on the infected host.
g. Re-dispersal / Dissemination of spores.
h. Perennating stages.
i. Inoculum potential and density in the seed, soil and air.

3. Environmental factors

a. Temperature.
b. Humidity.
c. Light intensity.
d. Wind velocity.

Methods of disease forecasting

1. Forecasting based on correlative information

- Weather data of numerous years are collected and correlated with the intensity of the diseases.
- The data are compared and then the forecasting of the disease is done.
- Forecasting criteria developed from comparisons of disease observation with standard meteorological data have been provided for diseases like fire blight of apple and powdery mildew of barley.

2. Use of computer for disease forecasting

- Forecasting of disease is made by the use of computers.
- The computers are fed with weather data collected on the farm by individual farmers.
- The computer then processes the data, determines whether an infection period is imminent, likely to occur or cannot occur, and a recommendation is made to the farmer as to whether or not to spray and what materials to apply.
- This system gives the results quickly.
- One such computer based programmes in the USA is known as 'Blitecast' for potato late blight.

Dutch rules

a. Late blight of potato

- Late blight of potato had been instrumental in development of diverse fundamental concepts in plant pathology and disease forecasting is no exception.
- Van Everdingen postulated Dutch rules in 1926.
- Followings are four conditions correlated with weather that could predict about appearance of late blight of potato in Netherlands:
    - Night temperature below dew point at least 4 hours.
    - Mainimum temperature of 10^0C or slightly above.
    - Clouds on the next day.
    - Rainfall during next 24 hours of at least 0.1 mm.

Notable contributions in late blight forecasting

1. Van Everdingen postulated Dutch rules in 1926.
2. Krause, *et al.* (1970), BLITECAST.
3. Singh, *et al.* (1980), JHULSACAST for Indo Gangetic plains.
4. Fry and Apple (1983), Integrated host resistance and fungicide weathering in BLITECAST.
5. Dommermuth (1988), Phytoprog I. late blight warning service.
6. Runno and Kopple (2002), NEGFRY.
7. Grunwald (2002), Modified and validated SIMCAST.

EPIPRE (EPIdemics PREdiction and PREvention)

- System of supervised control of diseases and pests in winter wheat.
- Participating farmers do their own disease and pest monitoring, simple and reliable observation and sampling techniques.

- Farmers send their field observations to the central team, which enters them in the data bank.
- Field data are updated daily by means of simplified simulation models.
- Expected damage and loss are calculated and used in a decision system, that leads to one of three major decisions
  - Treat
  - Don't treat
  - Make another field observation
  - The start of EPIPRE in 1978 was promoted by the heavy epidemics of yellow rust in 1975 and 1977.

Empirical vs Fundamental forecasting

- Bourke (1955) designated two methods of disease forecasting.
- They are 'empirical' and 'fundamental'

Empirical methods

- Method in which correlation between the results of disease surveys and the corresponding weather conditions in a particular area or near the same area has to be related to the biology of the host plant and pathogen.

Fundamental methods

- Method which uses results obtained from laboratory research on weather factors affecting pathogen, host, and the disease process as the starting point, interpreted and applied according to conditions in the region concerned.
  - Neither method is more reliable than the other.
  - Both need to be tested and evaluated in actual use.

Uses of Disease Forecasts

1. Timely plant protection.
2. Loss assessment.
3. Making strategic decision.
   - Prediction of the risks involved in planting a certain crop.
   - Deciding about the need to apply strategic control measures (soil treatment, planting a resistant cultivar, etc.)

4. Making tactical decision

- To decide about the need to implement disease management measure.
- To develop systems that help growers make economic decisions for managing diseases.
- To support a producers decision making process for determining cost and benefits for applying pesticides, selecting seed or propagation materials, or whether to plant a crop in a particular area.

Successful Plant Disease Forecasting System

- **Simplicity** - Simpler the system, the more likely it will be applied and used by producers.
- **Importance** -Disease is of economic importance to the crop.
- **Usefulness** -Forecasting model should be applied when the disease and/or pathogen can be detected reliably.
- **Reliability** -Use of sound biological and environmental data.
- **Availability** –Essential information about the components of the disease triangle should be available.
- **Multipurpose applicability** -Monitoring and decision-making tools for several diseases and pests should be available.
- **Cost effectiveness** -Forecasting system should be cost affordable relative to available disease management tactics.

Success of a Forecasting

Success of a forecasting system depends on

- Commonness of epidemics.
- Accuracy of predictions of epidemic risk
- Ability to deliver predictions in a timely fashion.
- Ability to implement a control tactic.
- Economic impact of using a predictive system.

6

Expert Systems in Plant Pathology

- Expert system are computer programs that emulate the logic and problem solving proficiency of human expert.
- Expert systems are the frontiers, which combines and integrate both science of plant pathology, art of diagnosis and disease management.
- It is one of the most efficient extension tools to take the technology from scientists to the farmers directly without any dilution of content which normally creeps in because of the number of agencies involved in normal technology transfer systems.
- An Agricultural Expert System is a Decision Support System for Agricultural Extension Agents who have to decide what advice to be offered to farmers who in turn decide about suitable action based on it.
- These computer decision support systems were formally introduced in Plant Pathology in 1987.
- The computerized disease forecasting systems of 1970s (BLITECAST) and 1980s (apple scab forecaster) were the precursors to expert system.
- In principle the Expert system are an artificial intelligence application that uses encoded knowledge from a human expert.
- Expert system has the ability to use complete or incomplete data; accordingly it will assign certainty values to the solutions of the problems.
- The Expert system are programmed to review a consultation and provide user with an explanation.
- Three main areas involved in decision making are
  - Pest risk estimation.
  - Current threat assessment.
  - Pesticide application.

Expert System used in Plant Pathology

Expert System	Description
PLANT/ds	First expert system (1983) developed for diagnosis of 17 soybean diseases (USA)
EPIPRE	Epidemic prediction and prevention for spring and winter wheat
POTATOES	Late and Early blight of potato
TOMEX	Diagnosis of 37 diseases of tomato
POMME	For management of disease and insects of apple
PSACO	For management of diseases and insects of apple
PRO-PLANT	For cereals, potato, sugarbeet and vegetables
More Crop	For management of rusts and other problems
TOMEx –UFV	For diagnosis of tomato diseases
SIMPHYT-III	For *Phytophthora infestans*

7

Modern Tools in Epidemiology

I. Study of plant disease epidemiology facilitated greatly by new methods and new equipment.

II. This make possible studies of aspects of plant disease that were impossible or very difficult to study earlier.

III. Some of the methods and other equipment that have been used to great advantage in plant disease epidemiology include the following.

- Molecular Tools
- Geographic Information System
- Global Positioning System
- Remote Sensing
- Geostatistics
- Image Analysis
- Information Technology

Molecular Tools

1. Development and use of genetic (DNA) probes that allow the definitive detection and identification of a plant pathogen
   - Within or on the surface of a plant tissue.
   - In a mixture with other microorganisms.
   - Even in the vicinity of the host plant.
2. Detection and identification of a pathogen by its genetic probe are made immensely more effective through the use of the PCR technique.
3. PCR amplifies greatly a specific fragment of DNA present on a probe and produces millions of copies of it.
4. These copies are then abundant enough to be detected, identified, and studied by conventional or other molecular tools.
5. RAPD markers are often used to detect genetic similarities among pathogenic strains known to show genetic heterogeneity.

6. It can also be used easily for designing sequence characterized amplified region (SCAR) markers for detecting the pathogen in infected plant tissue.

Significance of Molecular Techniques

- To detect pathogen arrival much earlier than could be detected before.
- To thereby allowing the grower time to get ready.
- To apply whatever management treatment is most effective against the pathogen.
- Detect any new mutant pathogens early.
- Tolerate the fungicide to which they were sensitive before and thus produce a new resistant race.

Detection of such changes in pathogens is of paramount importance in epidemiology because such changes in pathogens make useless and necessitate immediate revision of any previous predictions about the development of the epidemic and recommendations for management of the disease.

Geographic Information System (GIS)

1. Computer system that can be installed on any recent model desktop computer.
2. Capable of assembling, storing, manipulating, and displaying data that are referenced by geographic coordinates.
3. GIS is adaptable to operations of any size, and data can be used at any scale from a single field to an agricultural region.
4. It is used to better understanding and management of plant disease epidemics and manage the environment.
5. GIS techniques allow one to make connections between events based on geographic proximity, connections that are essential to the understanding and management of epidemics but which often go unrecognized without GIS.
6. GIS techniques can even incorporate disease forecasting systems, although the time and cost for it may be prohibitive.
7. GIS is often used for the spatial and temporal analysis of disease development over relatively large geographic areas.
8. It helps determine the role and relative importance of various parts of these areas in the initiation and development of the epidemic.

Global Positioning System (GPS)

1. GPS consists of a hand held device that is coordinated with a global system of man-made satellites.
2. Depending on the accuracy and coordination, it provides quite accurate readings of the coordinates of the position of the device.
3. GPS enables one to pinpoint an individual tree or a specific area or areas of the field that are affected by a pathogen, which then can be visited and examined again periodically for incremental advance of the symptoms.
4. Similarly, the selected trees or areas could be treated with the appropriate pesticide or other treatment wherever the pathogen is present without the need to treat the entire field.
5. GPS can also be used to apply pesticides, plant nutrients, and so on in only the areas of the field that are infested with the pathogen or in areas deficient in a particular micro- or macronutrient.
6. Elimination of the pathogen from the field by early detection and treatment is often effective in not allowing the pathogen to cause an epidemic in the field and beyond.

Geostatistics

1. Consist of various "geostatistical" techniques that are applied in plant disease epidemiology.
2. It characterize quantitatively spatial patterns of disease development or the development of pathogen populations in space and over time.
3. These techniques have the capability to take into account the characteristics of spatially distributed variables whether they are random or systematic.
4. Can also be used for studying continuous and discrete variables.
5. It do not require as exacting asumptions of stationarity as do other spatial autocorrelation techniques.
6. The spatial dependence or connection can be analyzed with semivariograms.
7. The latter quantify spatial dependence by determining the variation between samples.

Remote Sensing

1. Refers to the use of instruments for measuring electromagnetic radiation reflected or emitted from an object.
2. Instruments record reflected or emitted radiation in the ultraviolet, visible, or infrared part of the spectrum.

3. Instruments used for remote sensing may be hand-held, ground-based cameras with films and filters, digital cameras, video systems, and radiometers or they may be carried on balloons, aircraft, and satellites.
4. Various remote sensing instruments store data obtained from field situations, and data are then printed out and are analyzed directly or by transferring them to a computer and creating visual images of data

Image Analysis

1. Image analysis refers to photography and electronic image analysis, usually of large areas of fields or of mountains.
2. Images or photographs are taken through aerial photography, ground-based sensor data, and satellite-borne and airborne sensors.
3. Airborne mul-tispectral scanning is studied and used widely for the surveillance of plant diseases, pests, and environmental stresses in agriculture.
4. Often, infrared light or light of other wavelengths is used for the detection of the onset and progress of a plant disease among the crop plants in the field or among the fruit or forest trees in a mountain.
5. Plants and trees, when infected with various pathogens or subjected to other stresses, turn light green, then chlorotic (yellowish), and then brown and have different reflectances from the healthy plant.
6. These colors or shades of colors become more distinct when photographed with the wavelengths mentioned previously than when photographed with the normal visible light spectrum.
7. More importantly, such photographs can be examined and analyzed with specific equipment that not only better distinguishes such disease-discolored plants, but can also provide a count of the newly infected plants as well as measure the changes in intensity of the images of previously diseased plants.
8. Image analysis can provide a measure of the severity of the disease in each plant or area of infected plants and, by repetition of the photography at regular intervals, provide a measure of the rate of progress of the disease.

Information Technology

1. Use of computers alone or in combination with other electronic devises.
2. They help collect data on plant diseases at various levels and various locations in a continuous manner.

3. Data are either stored or are organized, integrated, and analyzed in tremendous quantities and at thus far unimaginable speeds and eventually are used to produce visual images or written reports and recommendations.
4. Electronic information technology can, above all, describe and display spatial patterns of characteristics of different pathogens, such as their genotypes, at the scale of an agricultural region.

8

Prognosis

- Prognosis is characterized as the prediction of the outbreak, development and outcome of disease.
- Objective is to decide in advance whether expected damage is threatening and whether control measures are to be taken.
- Prognosis has been attempted at two levels:

  a) Date prognosis

  b) Loss prognosis

Types of Prognosis

a) Date prognosis

- Mainly deals with the prediction of the outbreak of a disease.
- Following two criteria are considered for date prognosis:
  - Estimate the initial population of pathogens and their antagonist, if feasible.
  - Record weather parameters and their prediction over prognosis time span.

b) Loss prognosis

- Based on estimate of expected economic loss, in relation to disease intensity-crop damage.
- Constitutes the basis or the decision as to spray or not.
- In loss prognosis in addition to economic aspects, consideration of the biological and eco toxicological consequences of a plant protection measure are important factors.

Unit 8: Phanerogamic Parasites and Non-parasitic Diseases

1

Non-Parasite Diseases

Introduction

1. Plants grow best within certain ranges of the various abiotic factors that make up their environment.
2. Such factors include temperature, soil moisture, soil nutrients, light, air and soil pollutants, air humidity, soil structure, and pH.
3. Although these factors affect all plants growing in nature.
4. Their importance is considerably greater for cultivated plants, which are often grown in areas that are at the margins and beyond their normal habitat and, therefore, that barely meet the requirements for normal growth.

General Characteristics

1. Abiotic, i.e., noninfectious diseases of plants, is that they are caused by the lack or excess of something that supports life.
2. Occur in the absence of pathogens.
3. Therefore, cannot be transmitted from diseased to healthy plants.
4. May affect plants in all stages of their lives (e.g., seed, seedling, mature plant, or fruit).
5. They may cause damage in the field, in storage, or at the market.
6. Symptoms may range from slight to severe, and affected plants may even die.

Diagnosis

1. Diagnosis of noninfectious diseases is sometimes made easy by the presence of characteristic symptoms known to be caused by the lack or excess of a particular factor on the plant.
2. At other times, diagnosis can be arrived at by carefully examining and analyzing several factors:
   - Weather conditions prevailing before and during the appearance of the disease.

- Recent changes in the atmospheric and soil contaminants at or near the area where the plants are growing.
- Cultural practices, or possible accidents in the course of these practices, preceding the appearance of the disease.

3. Often, however, the symptoms of several noninfectious diseases are too indistinct and closely resemble those caused by several viruses, mollicutes, and many root pathogens.
4. Diagnosis of such noninfectious disease then becomes a great deal more complicated.
   - One must obtain proof of absence from the plant of any of the pathogens that could cause the disease.
   - One must reproduce the disease on healthy plants after subjecting them to conditions similar to those thought of as the cause of the disease.
   - To distinguish further among environmental factors causing similar symptoms.
   - Investigator must cure the diseased plants.
   - if possible, by growing them under conditions in which the degree or the amount of the suspected environmental factor involved has been adjusted to normal.

Temperature Effects

- Plants normally grow at a temperature range from 1 to 40°C.
- Most kinds of plants growing best between 15 and 30°C.
- Perennial plants and dormant organs (e.g., seeds and corms) of annual plants may survive temperatures considerably below or above the normal temperature range of 1 to 40°C.
- Young, growing tissues of most plants, however, and the entire growth of many annual plants are usually quite sensitive to temperatures near or beyond the extremes of this range.
- A plant may also differ in its ability to withstand extremes in temperature at different stages of its growth.
- Thus, older, hardened plants are more resistant to low temperatures than young seedlings.

High-Temperature Effects

1. Plants are generally injured faster and to a greater extent when temperatures become higher than the maximum for growth than when they are lower than the minimum.

2. High temperature seems to cause its effects on the plant in conjunction with the effects of other environmental factors, particularly excessive light, drought, lack of oxygen, or high winds accompanied by low relative humidity.
3. High temperatures are usually responsible for SUNSCALD INJURIES.
    - Appearing on the sun-exposed sides of fleshy fruits and vegetables.
    - Effected crops; Peppers, Apples, Tomatoes, Onion Bulbs, Potato Tubers.
    - On hot, sunny days the temperature of the fruit tissues beneath the surface facing the sun may be much higher than that of those on the shaded side and of the surrounding air.
    - This results in discoloration, a water-soaked appearance, blistering, and desiccation of the tissues beneath the skin, which leads to sunken areas on the fruit surface.
4. Too high a soil temperature at the soil line sometimes kills young seedlings or causes cankers at the crown on the stems of older plants.
5. High tempertures also seem to be involved in the water core disorder of apples and, in combination with reduced oxygen, in the blackheart of potatoes.

Low-Temperature Effects

1. Far greater damage to crops is caused by low than by high temperatures.
2. They may also cause excessive sweetening and, on frying, undesirable caramelization of potatoes due to the hydrolysis of starch to sugars at the low temperatures.
3. Temperatures below freezing cause a variety of injuries to plants. Such injuries include
    - Damage caused by late frosts to young leaves and meristematic tips.
    - Damage of entire herbaceous plants.
    - Frost killing of buds of peach, cherry, and other trees.
    - Killing of flowers, young fruit, and sometimes, succulent twigs of most trees.
4. Fleshy tissues, such as tomato fruit, canola pods, and potato tubers, may be injured at subfreezing temperatures.
5. In potatoes, the injury varies depending on the degree of temperature drop and the duration of the low temperature.

6. Early injury affects only the main vascular tissues and appears as a ring-like necrosis; injury of the finer vascular elements that are interspersed in the tuber gives the appearance of net-like necrosis.

Mechanisms of Low- and High-Temperature Injury to Plants

1. **High temperatures**
    - Apparently inactivate certain enzyme systems and accelerate others, thus leading to abnormal biochemical reactions and cell death.
    - Cause coagulation and denaturation of proteins, disruption of cytoplasmic membranes, suffocation, and possibly release of toxic products into the cell.
2. **Low temperatures**

- Injure plants primarily by inducing ice formation between or within the cells.
- when the intercellular water becomes ice, more vapor (water) moves out of the cells and into the intercellular spaces, where it also becomes ice.
    - The reduced water content of the cells depresses further the freezing point of the intracellular water.
    - Ice crystals form within the cell, disrupt the plasma membrane, and cause injury and death to the cell.
    - Ice formation in super cooled water within leaves is influenced greatly by the kinds and numbers of epiphytic bacteria that may be present on the leaves.
    - Certain strains of some pathogenic (e.g., *Pseudomonas syringae)* bacteria and of some saprophytic bacteria, when present on or in the substomatal cavities of leaves, act as catalysts for ice nucleation.
    - By their presence alone, such ice nucleation-active bacteria induce the supercooled water around them and in the leaf cells to form crystals, thereby causing frost injury to the leaves, blossoms, and so on at temperatures considerably higher (-1°C) than would have happened in the absence of such bacteria (approximately -5 to -10°C).
    - This loss causes plasmolysis and dehydration of the protoplasm, which may cause coagulation.

Moisture Effects

Low Soil Moisture Effects

1. Moisture disturbances in the soil are probably responsible for more plants growing poorly and being unproductive annually, over large areas, than any other single environmental factor.
2. Small or large territories may suffer from drought over time.
3. The subnormal amounts of water available to plants in these areas may result in reduced growth, a diseased appearance, or even death of the plants.
4. Lack of moisture may result in patches of diseased looking plants, while the immediately surrounding areas appear to contain sufficient amounts of moisture and the plants in them grow normally.
5. Plants suffering from lack of sufficient soil moisture usually remain stunted, are pale green to light yellow, have few, small and drooping leaves, flower and fruit sparingly, and, if the drought continues, wilt and die .
6. Annual plants are considerably more susceptible to short periods of insufficient moisture.
7. Even perennial plants and trees are damaged by prolonged periods of drought and produce less growth, small, scorched leaves and short twigs, dieback, defoliation, and finally wilting and death.
8. Plants weakened by drought are also more susceptible to certain pathogens and insects.

Low Relative Humidity Effects

1. Lack of moisture in the atmosphere, i.e., low relative humidity, is usually temporary and seldom causes damage.
2. When combined with high wind velocity and high temperature, however, it may lead to excessive loss of water from the foliage.
3. It may result in leaf scorching or burning, shriveled fruit, and temporary or permanent wilting of plants.
4. Conditions of low relative humidity are particularly common and injurious to house plants during the winter.
5. In modern homes and apartments, heating provides comfortable temperatures for plant growth, but it often dries the air to relative humidities of 15 to 25% which are equivalent to that of desert environments.

6. Potted plants kept under these conditions not only use up the water much faster, grow poorly, and may begin to wilt sooner, but the leaves, especially the lower ones, of many kinds of plants become spotted or scorched and fall prematurely, while their flowers suddenly wither and drop off.
7. These effects are particularly noticeable when plants are brought into such a hot, dry house directly from a cool, moist greenhouse or florist's shop.

High Soil Moisture Effects

1. Excessive soil moisture occurs much less often than drought where plants are grown.
2. Poor drainage or flooding of planted fields, gardens, or potted plants may result in more serious and quicker damage, or death, to plants than that from lack of moisture.
3. Poor drainage results in plants that lack vigor, wilt frequently, and have leaves that are pale green or yellowish green.
4. Flooding during the growth season may cause permanent wilting and death of succulent annuals within 2 to 3 days.
5. Excessive soil moisture caused by flooding or by poor drainage, the fibrous roots of plants decay, probably because of the reduced supply of oxygen to the roots
6. Oxygen deprivation causes stress, asphyxiation, and collapse of many root cells.
7. Wet, anaerobic conditions favor the growth of anaerobic microorganisms that, during their life processes, form substances, such as nitrites, that are toxic to plants.
   - Common symptom of houseplants, and sometimes of outdoor plants.
   - Caused by excessive moisture is the so-called edema [or oedema (swelling)].
   - Edema appears as numerous small bumps on the lower side of leaves or on stems.

EDEMA

1. Common symptom of houseplants, and sometimes of outdoor plants.
2. Caused by excessive moisture is the so-called edema [or oedema (swelling)].

3. Appears as numerous small bumps on the lower side of leaves or on stems.
4. The "bumps" are small masses of cells that divide, expand, and break out of the normal leaf surface and at first form greenish- white swellings or galls.
5. Later, the exposed surface of the swelling becomes rusty colored and has a corky texture.
6. Edema is caused by overwatering, especially during cloudy, humid weather, and can be avoided by reduced watering and providing better lighting and air circulation to the plant.
7. Many other disorders are caused by excessive or irregular watering.
8. Example, that tomatoes and some other fruits, such as cherries and grapes, grown under rather low moisture conditions at the time they are ripening often crack if they are suddenly supplied with abundant moisture by overwatering or by a heavy rainfall.
9. Bitter pit of apples, consisting of small, sunken, blackspots on the fruit, is the result of an irregular supply of moisture, although excessive nitrogen and low calcium fertilization also seem to be involved in bitter pit development.

Inadequate Oxygen

1. Low oxygen conditions in nature are generally associated with high soil moisture or high temperatures.
2. A combination of high soil moisture and high soil or air temperature causes root collapse in plants.
3. The first condition, apparently, reduces the amount of oxygen available to the roots, whereas the other increases the amount of oxygen required by the plants.
4. The two effects together result in an extreme lack of oxygen in the roots and cause their collapse and death.
5. Best known such case is the development of the so-called

Black heart of Potato

- Fairly high temperatures stimulate respiration and abnormal enzymatic reactions in the potato tuber.
- Oxygen supply to the cells in the interior of the tuber is insufficient to sustain the increased respiration and the cells die of suboxidation.
- Enzymatic reactions activated by the high temperature and suboxidation go on before, during, and after the death of the cells.

- These reactions abnormally oxidize normal plant constituents into dark melanin pigments.
- The pigments spread into the surrounding tuber tissues and finally make them appear black.

Light

I. Lack of sufficient light

- Retards chlorophyll formation and promotes slender growth with long internodes.
- Leading to pale green leaves, spindly growth, and premature drop of leaves and flowers.
- This condition is known as ETIOLATION.
- Etiolated plant develop only when plants are spaced too close together or when they are growing under trees or other objects.
- Etiolation of various degrees, where plants often receive inadequate light.
- Etiolated plants are usually thin and tall and are susceptible to lodging.

II. Excess light

- Rather rare in nature and seldom injures plants.
- Many injuries attributed to light are probably the result of high temperatures accompanying high-light intensities.
- Excessive light cause SUNSCALD of pods of beans grown at high altitudes.
- It is due to the absence of dust, more light of short wavelengths reaches the earth.
- The pods develop small, water-soaked spots that quickly become brown or reddish brown and shrink.

Nutritional Deficiencies in Plants

1. Plants require several mineral elements for normal growth.
2. Some elements, such as **nitrogen, phosphorus, potassium, calcium, magnesium, and sulfur,** needed in relatively large amounts, are called MAJOR ELEMENTS,
3. Element such **as iron, boron, manganese, zinc, copper, molybdenum, and chlorine**, needed in very small amounts, are called trace or minor elements or MICRONUTRIENTS.
4. Both major and trace elements are essential to the plant.

5. When they are present in the plant in amounts smaller than the minimum levels required for normal plant growth, the plant becomes diseased and exhibits various external and internal symptoms.
6. The symptoms may appear on any or all organs of the plant, including leaves, stems, roots, flowers, fruits, and seeds.

Nutritional Deficiencies or Disorder in Plants

1. Deficiency of minerals *viz, Nitrogen, Phosphorus, potash, manganese , zinc, copper , iron, magnesium, boron,* , etc. results in disorders in plant metabolism and cause *hunger signs* in the crops
2. Excess and deficiency of minerals disturbs nutritional balance needed for good metabolism in the plant, hence hindering the effect of essential element.
3. Deficiencies and excess of minerals also reduce the resistance of plant to fungal, bacterial and other diseases.
4. Kind of symptoms produced by deficiency of a certain nutrient depend primarily on the functions of that particular element in the plant.
5. Nutrient mobility plays an important role in determining the site of the deficiency symptoms.
6. For *highly mobile* elements like *nitrogen and potassium*, the deficiency symptoms appear predominantly in *older* and mature leaves.
7. Deficiency of *poorly mobile* elements, like *calcium, boron* and *iron* show up in *younger* leaves.
8. The symptoms of deficiency of *moderately mobile* elements like *magnesium* and *sulphur*, are uniformly spread all over the plant.

Plant disease due to lack of minerals

S.No.	Deficient Nutrient	Disease
1	Nitrogen (N)	Red leaf of cotton
2	Phosphorus (P)	Dwarfening of cotton
3	Potassium (K)	Cotton rust leaf spot of alfa alfa.
4	Magnesium (Mg)	Sand drown disease in tobacco
5	Boron (B)	Internal cork of apple Cracked stem of celery Black tip of mango Heart rot of sugarbeet Brown heart of cabbage and turnip Internal brown spot of sweet potato Terminal bud breakdown of tobacco Fruit pitting and dieback of olive Hollow stem of brassica
6	Copper (Cu)	Reclamation disease of oats Diebacks of citrus Wither tip of apple
7	Manganese(Mn)	Pahala blight of sugarcane Marsh spot of garden pea Grey speck of oats
8	Zinc (Zn)	White bud of corn Little leaf of apple Bronzing of twigs Rosette of fruits trees Mottle leaf of citrus Khaira disease of rice
9	Iron (Fe)	Green netting of citrus
10	Molybdenum (Mo)	Whiptail disease of cauliflower.
11	Calcium(ca)	Blossom end rot of tomato Black heart of celery Wither tip of flax

2

Phanerogamic Parasites

Plant Diseases caused by Parasitic higher plants

1. More than 2500 species of higher plants are known to live parasitically on other plants.
2. Their main common characteristic is that these parasites are
   - Vascular plants that have developed specialized organs which penetrate the tissues of other (host) vascular plants.
   - Establish connections to the host plant vascular elements.
   - And absorb nutrients from them.
3. These parasitic plants produce flowers and seeds and belong to several widely separated botanical families.
4. They vary greatly in their dependence on their host plants.
   - Some, e.g, Mistletoes, have chlorophyll but no roots so they depend on their hosts only for water and minerals.
   - Others, e.g., Dodder, have little or no chlorophyll and no true roots so they depend entirely on their hosts for their existence.
5. Relatively few of the known parasitic higher plants cause important diseases on agricultural crops or forest trees.

Classification of Flowering Plant Parasites

i) Complete Parasite (Holoparasites)

a. Root: Rafflesia, Orobanchea (Broom rape)

b. Stem: Cuscuta (Dodder. Amarvel)

ii) Partial Parasites (Semi Parasites)

a. Root: Striga (Sandle wood, witch weed)

b. Stem : Loranthus (Banda or Deudrophae)

Most Common and Serious Parasites

Cuscutaceae

1. Genus: Cuscuta, the dodders of alfalfa, onion, potato, and numerous other plants.

Lauraceae

1. Genus: Cassytha, *C. filiformis* infecting shrubs and trees in the Caribbean islands and in Florida

Viscaceae

1. Genus: Arceuthobium, the dwarf mistletoes of conifers.
   Phoradendron, the American true mistletoes of broad-leaved trees.
   Viscum, the European true misteltoes.

Orobanchaceae

1. Genus: Orobanche, the broomrapes of legumes, solanaceous, and other plants.

Scrophulariaceae

1. Genus: Striga, the witchweeds of many monocotyledonous and some legume plants.

I. Dodders and the dwarf and true mistletoes

- Attach themselves to and parasitize aboveground parts, i.e., shoots and branches of their hosts.
- Cause relatively small economic losses.

II. Witch weeds and the broomrapes

- Attach themselves to and parasitize the roots of their host plants and cause serious economic losses.
- Witchweeds are one of the biggest biological hindrances in grain and corn production in Africa.

Parasitic Higher Plants: Dodder

1. Crops that suffer losses from dodder include alfalfa, onions, sugar beets, several ornamentals, and potatoes.
2. Dodder affects the growth and yield of infected plants.
3. Losses range from slight to complete destruction of the crop in the infested areas.
4. Names such as strangle weed, pull-down and hellbind, by which dodder is referred to in different areas, are descriptive of the ways in which dodder affects its host plants.

5. Dodder may also serve as a bridge for transmission of viruses from virus infected to virus-free plants as long as both plants are infected by the same dodder plant.

Symptoms

1. Orange or yellow vine strands grow and entwine around the stems and the other aboveground parts of the plants.
2. The growing tips reach out and attack adjacent plants until a circle of infestation, up to 10 feet in diameter, is formed by a single dodder plant.
3. Dodder-infested patches in the field continue to enlarge during the growth season and, in perennial plants such as alfalfa, become larger every year.
4. During late spring and in the summer, dodder produces massed clusters of white, pink, or yellowish flowers, which soon form seed.
5. Infected host plants become weakened by the parasite, their vigor declines, and they produce poor yields.
6. Many are smothered and may be killed by the parasite.
7. As the infection spreads, several patches coalesce and form large areas covered by the yellowish vine of the parasite.

The Pathogen: *Cuscuta* spp

- Slender, twining plant.
- Stem is tough, curling, threadlike, and leafless, bearing only minute scales in place of leaves.
- Stem is usually yellowish or orange in color, sometimes tinged with red or purple; sometimes it is almost white.
- Clusters of tiny flowers occur on the stem from early June until frost.
- Gray to brown seeds are produced in abundance by the flowers and mature within a few weeks after bloom.

Development of Disease

1. Seed overwinters in infested fields or is mixed with the seed of crop plants.
2. During the growing season the seed germinates and produces a slender yellowish shoot but no roots.
3. This leafless shoot rotates as though in search of a host.
4. Stems in contact with a susceptible host encircle the host plant, send haustoria into it, and begin to climb the plant.
5. Haustoria penetrate the stem or leaf, and reach into the vascular tissues, from which they absorb foodstuffs and water.

6. Soon after contact with the host is established, the base of the dodder shrivels and dries so that the dodder loses all connection with the ground and becomes completely dependent on the host for nutrients and water.
7. Dodder continues to grow and expand, and its twisting tips reach out and attack adjacent plants, forming patches of infected plants.
8. Growth of infected plants is suppressed and they may finally die.
9. In the meantime, the dodder plant has developed flowers and produced seeds.
10. Seeds fall to the ground where they either germinate immediately or remain dormant until the next season.
11. Seed may be spread to nearby areas by animals, water, and equipment, and over long distances by contaminated crop seed.

Control

1. By frequent tillage and flaming.
2. Use of dodder-free seed.
3. By cleaning equipment thoroughly before moving it from dodder-infested fields to new areas.
4. Limiting the movement of domestic animals from infested to dodder free fields.
5. If dodder is already present in the field, scattered patches may be sprayed early in the season with contact herbicides.
6. Cutting or burning of patches, kills both the dodder and the host plants but prevents dodder from spreading and from producing seed.

Witch Weed

- In 1956, the weed was discovered for the first time in America, in North and South Carolina.
- It attacks mostly monocots such as corn sorghum, millet, upland rice, and sugarcane, but also cowpeas, peanuts, other legumes (, sweet potato, and tobacco. Infected plants become stunted and chlorotic.
- Heavily infected plants usually wilt and die.
- Losses vary and may range from slight to 100%.

Symptoms

- Affected plants remain stunted, wilt, and turn yellowish.
- Death may follow these symptoms if the plants are heavily parasitized.
- Infected roots bear a large number of witchweed haustoria, which are attached to the root and feed on it.

The Pathogen: *Striga* spp

- Witchweed is a small, pretty plant.
- It has a bright green, slightly hairy stem and leaves and grows 15 to 30centimeters high.
- Produces many branches both near the ground and higher on the plant.
- Flowers are small and usually red or yellowish, or white, always having yellow centers.
- After pollination, seed pods or capsules develop, each containing more than a thousand tiny brown seeds.
- A single plant may produce from 50,000 to 500,000 seeds.
- Root of witchweed is white and round in cross section.
- It has no root hairs, for it obtains all nutrients from the host plant through haustoria.

Development of Disease

- Parasite overwinters as seeds, most of which require a rest period of 15 to 18 months before germination.
- Although some can germinate without any dormancy.
- Seeds close to host roots germinate and grow toward these roots, attracted by the exudates of the host roots.
- As soon as the witch weed rootlet comes in contact with the host root, its tip swells into a bulb-shaped haustorium.
- Haustorium dissolves and penetrates the host roots within 8 to 24 hours and advances into the roots.
- Finally, its leading cells, usually tracheids, reach the vessels of the host roots.
- Tracheids force their way into the vessel, from which they absorb water and nutrients.
- Although the chlorophyll of witchweed plants is functional, manufactured foodstuffs still move from the host plant into the parasite.
- Disease spreads in the field in a circular pattern.
- Circle of infected plants increases year after year as the witchweed seeds spread in increasingly larger areas.
- Seeds are spread by wind, by water, by contaminated tools and equipment, or by contaminated soil carried on farm machinery.

Control

- Introduction of witchweed should be avoided by all means.
- Catch crops, consisting of host plants, may be planted to force the germination of witchweed seed, and the witchweed plants then can be destroyed by plowing under or by the use of herbicides.
- Trap crops, consisting mostly of nonhost legumes, may be used to stimulate the germination of witchweed seeds, which, however, cannot infect the trap plants and therefore starve to death.
- Use of resistant cultivars.
- Seed treatment with herbicides of differential toxicity.
- Use of witchweed-infecting fungi as a biological control are possible control
- Usually, a combination ofthe aforementioned methods is required to prevent witchweed plants from flowering and seeding.

Broomrapes

- Broomrapes occur in warm and dry regions worldwide.
- They attack several hundred species of herbaceous dicotyledonous crop plants.
- In some areas, broomrapes cause losses varying from 10 to 70% of the crop.

Symptoms

1. Occur in small, patches and may be stunted to various degrees, depending on how early in their lives and by how many broomrapes they were infected.
2. Orobanche sp., is a whitish to yellowish-brown annual plant 15 to 50 centimeters tall.
3. It has a fleshy stem and scale-like leaves and produces numerous pretty, white, yellow-white, or slightly purple, snapdragon-like flowers arising singly along the stem.
4. The broomrapes produce seed pods about 5 millimeters long, each containing several hundred minute seeds.

Development of Disease

- Broomrapes overwinter as seeds, which may survive in the soil for more than 10 years.
- Seeds germinate only when roots of certain plants grow near them, although not all these plants are susceptible to the pathogen.

- On germination the seed produces a radicle, which grows toward the root of the host plant, becomes attached to it, and produces a shallow cup-like appressorium that surrounds the root.
- From the appressorium, a mass of undifferentiated cells penetrate the host, extend to and, occasionally, into the xylem, and absorb nutrients and water from it.
- Some of these cells differentiate into parasite xylem vessel elements and connect the host xylem, with the main vascular system of the parasite.
- Other undifferentiated cells become attached to phloem cells and obtain nutrients from them, which they transport back to the parasite.
- Soon the parasite begins to develop a stem, which appears above the soil line and looks like an asparagus shoot.
- Meanwhile, the original root produces secondary roots that grow outward until they come in contact with other host roots to which they become attached and subsequently infect.
- The broomrape stems continue to grow and produce flowers and seeds, which mature and are scattered over the ground in less than two months from the emergence of the stems.

Loranthus or Bandgul

- Loranthus is common parasites of mango trees.
- The parasite attacks, the aerial parts of the host trees, by developing haustoria and obtain its nourishment directly from the vascular system of the host plant.
- Since the parasite attacks the aerial part of host tree, situated far above the soil level and as such devoid of root system of its own.
- The fruit is fleshy and contains a solitary seed. It is sweet eaten by birds and animals.
- The parasite is spread by dispersal of its seed, mostly through birds and to some extent by other animals.
- The damage done by the parasite is most marked in production of new growth of the host. Leaves are reduced in size and show unhealthy green colour. The quantity and yield of fruit is considerably lowered.

Management

- Injection of $CuSO_4$ or 2-4-D into the infected branches.
- Scrapping of parasite before seedling from the infected bunches.
- Sowing of branches sufficient low to the tumours.

Parasitic Green Algae

- Algae are the organisms that can carry on photosynthesis.
- Algae are sometimes considered as protists with chloroplasts.
- There are eight groups of algal protists.
- Some algae, the so-called blue-green cyanobacteria, belong to the kingdom Eubacterial Prokaryotes.
- Algae are the main producers of photosynthetic materials in aquatic ecosystems.
- Green algae are single-celled organisms that form colonies, or multicellular, free-living organisms, all of which have chlorophyll b.
- Several algae are pathogens of other organisms.
- For example, cyanobacteria cause the black band disease that leads to bleaching and death of coral symbionts of the algae.
- Most of the green algae are free-living organisms, but several of their genera live as endophytes of many hydrophytes to which they seem to cause little or no damage.
- A few genera of green algae, however, are parasitic on higher plants.
- Green algal genera of Rhodochytrium of the family Chlorococcaceae and the genus Phyllosiphon of the family Phyllosiphonaceae infect numerous weeds and a few cultivated plants of relatively minor economic importance.
- Green algae of the genus Cephaleuros of the family Tentepohliaceae are true parasites of many wild and cultivated plants and cause diseases of economic importance.

Cephaleuros: General Characteristics

- Cephaleuros green algae, especially the genus *Cephaleuros virescens,* cause leaf spots and spots on stems plants belonging to more than 200 species.
- These green algae also cause lesions on fruit but less frequently.
- Some of the economically most important plants attacked by green algae are tea, coffee, cacao, black pepper, citrus, and mango.
- Consist of a vegetative thallus that is disc-like and is composed of cells arranged symmetrically.
- The algal thallus produces filaments that grow mostly between the cuticle and the epidermis of host leaves but, under some conditions, the filaments also grow between the palisade and the mesophyll cells of leaves.

- Cephaleuros algae produce filaments on which zoosporangia are produced.
- They reproduce by means of zoospores in zoosporangia, which can be disseminated by wind, rain splashes, and wind-driven rain.
- Zoospores can infect new leaves, shoots, and fruit of plants.
- Infections are much more common at the end of the rainy season.
- Following infection, plant cells next to the invading thallus turn yellow, while nearby cells enlarge and divide.
- If the plants are under stress, the infecting thallus expands, while cells in tissues invaded earlier die and produce a lesion.
- There may be so many lesions produced on leaves and shoots that they almost cover the entire surface.

Unit 9: Management of Plant Diseases

1

Management of Plant Diseases

- Study of the symptoms, causes, and mechanisms of development of plant diseases has an extremely practical purpose:
- It allows for the development of methods to combat plant diseases.
- Management increases the quantity and improves the quality of plant products available for use.
- Methods of management vary considerably from one disease to another, depending on the
  - Kind of pathogen
  - Kind of host
  - Interaction of the pathogen and host .
  - And many other variables.
- Plants are generally treated as populations rather than as individuals in managing diseases.
- Certain hosts (particularly trees, ornamentals, and, sometimes, other virus-infected plants) may be treated individually.
- Management measures are generally aimed at saving the populations rather than a few individual plants.
- Most serious diseases of crop plants appear on a few plants in an area year after year, spread rapidly, and are difficult to cure after they have begun to develop.
- Almost all management methods are aimed at protecting plants from becoming diseased rather than at curing them after they have become diseased.
- Few infectious plant diseases can be managmed satisfactorily in the field by therapeutic means.

Management Methods

Various management methods can be classified as

1. Regulatory
2. Cultural

3. Biological
4. Physical
5. Chemical

1. **Regulatory management** aims at excluding a pathogen from a host or from a certain geographic area.
2. **Cultural management** aims at helping plants avoid contact with a pathogen, creating environmental conditions unfavorable to the pathogen or avoiding favorable ones, and eradicating or reducing the amount of a pathogen in a plant, a field, or an area.
3. Most **biological** and some **cultural control** methods aim at improving the resistance of the host or favoring microorganisms antagonistic to the pathogen. A new type of biological control involves the transfer of genetic material (DNA) into plants and the generation of transgenic plants that exhibit resistance to a certain disease(s).
4. Finally, **physical** and **chemical** methods aim at protecting the plants from pathogen inoculum that has arrived, or is likely to arrive, or curing an infection that is already in progress.
5. Some recent (1995) and still mostly experimental chemicals operate by activating the defenses of the plant (systemic acquired resistance) against pathogens.

Sl.No	Methods	Aim
1	Cultural	• Helping plants avoid contact with a pathogen. • Creating environmental conditions unfavorable to the pathogen. • Eradicating or reducing the amount of a pathogen in a plant, a field, or an area. • Improving the resistance of the host.
2	Physical	• Protecting the plants from pathogen inoculum that has arrived, or is likely to arrive,. • Curing an infection that is already in progress.
3	Regulatory	• Aims at excluding a pathogen from a host or from a certain geographic area.
4	Biological	• Improving the resistance of the host. • Favoring microorganisms antagonistic to the pathogen. • Generation of transgenic plants that exhibit resistance to a certain disease(s).

5	Chemical	• Infecting the plants from pathogen inoculum that has arrived, or is likely to arrive. • Curing an infection that is already in progress. • Activating the defenses of the plant (systemic acquired resistance) against pathogens.

Effectiveness of Various Management Method

- Epidemiological studies and elucidating the development of diseases in an area over time, can help determine how effective various controls might be for a particular disease.

Monocyclic Pathogens

- Excluding or reducing the initial inoculum is most effective for the management of monocyclic pathogens.
- Controls such as crop rotation, removal of alternate hosts, and soil fumigation reduce the initial inoculum.

Polycyclic Pathogens

- With polycyclic pathogens, initial inoculum can be multiplied many times during the growing season.
- Reduction in the initial inoculum must usually be accompanied by another type of control measure (such as chemical protection or horizontal resistance) that also reduces the infection rate.
- Many controls, e.g., excluding a pathogen from an area, are useful for both monocyclic and polycyclic pathogen.

2

Physical Methods of Disease Management

Physical Control

Defintion

- Management of plant diseases using physical means is called physical control.
- Also called physical therapy or thermotherapy.
- It is mostly curative in nature

Principles

The Principles involved in physical control or thermotherapy is that the pathogens present in seed, soil, or planting material.

- Are inactivated or eliminated at temperatures nonlethal for the host tissues.
- Heat causes inactivation and not immobilization of the pathogen by heat.
- The rate at which the pathogen is inactivated is determined by temperature, the higher the temperature, the faster is the inactivation.

Physical Agents

- Physical agents used most commonly in controlling plant diseases.
- Most common physical gents used in plant disease management are
    - Temperature (high or low)
    - Dry air
    - Unfavorable light wavelengths
    - Various types of radiation
- Cultivation in glass or plastic greenhouses provides physical barriers to pathogens and their vectors and in that way protects the crop from some diseases.
- Plastic or net covering of row crops may protect the crop from infection by preventing pathogens or vectors from reaching the plants.

Different Methods of Physical Management

1. Disease management by heat treatment

- Soil sterilization by heat
- Soil solarization
- Hot-water treatment of propagative organs
- Hot-air treatment of storage organs

2. Disease management by eliminating certain light wavelengths

- Drying stored grains and fruit

3. Disease management by radiation

- Trench barriers against root-transmitted tree diseases

4. Disease management by refrigeration

Disease management by heat treatment

- Soil can be sterilized in greenhouses, in seed beds and cold frames.
- They can be sterilized by the heat carried in live or aerated steam or hot water.
- Soil is steam sterilized either in special containers (soil sterilizers), into which steam is supplied under pressure, or on the greenhouse benches, in which case steam is piped into and is allowed to diffuse through the soil.
- Most plant pathogenic fungi and bacteria, some worms, slugs, and centipedes, are usually killed at temperatures between 60 and 72°C.
- Nematodes, some oomycetes, and other water molds are killed, at about 50°C.
- Most weeds, plant pathogenic bacteria, plant viruses in plant debris, and most insects are killed at about 82°C
- Some plant viruses, such as tobacco mosaic virus (TMV), are killed at or near the boiling point, i.e., between 95 and 100°C.
- Soil sterilization is completed when the temperature in the coldest part of the soil has remained for at least 30 minutes at 82°C or above.
- All plant pathogens in the soil are killed at this temperature.
- Heat sterilization of soil can also be achieved by heat produced electrically rather than supplied by steam or hot water.
- Excessively high or prolonged high temperatures should be avoided during soil sterilization.
- Such condition leads to

  - Destroy all normal saprophytic microflora in the soil.
  - Release of toxic levels of some (e.g., Manganese) salts.
  - Accumulation of toxic levels of ammonia.
  - Which may damage or kill plants planted afterward.
  - Soil can be sterilized in greenhouses, in seed beds and cold frames.
- They can be sterilized by the heat carried in live or aerated steam or hot water.
- Soil is steam sterilized either in special containers (soil sterilizers), into which steam is supplied under pressure, or on the greenhouse benches, in which case steam is piped into and is allowed to diffuse through the soil.
- Most plant pathogenic fungi and bacteria, some worms, slugs, and centipedes, are usually killed at temperatures between 60 and 72°C.
- Nematodes, some oomycetes, and other water molds are killed, at about 50°C.
- Most weeds, the rest of the plant pathogenic bacteria, most plant viruses in plant debris, and most insects are killed at about 82°C.

Soil Solarization

- Clear polyethylene is placed over moist soil during sunny summer days.
- Temperature at the top 5 centimeters of soil may reach as high as 52°C compared to a maximum of 37°C in unmulched soil.
- Sunny weather needs to be continued for several days or weeks.
- Soil temperature increses from solar heat, known as solarization.
- It inactivates (kills) many soilborne pathogen fungi, nematodes, and bacteria near the soil surface.
- Ultimtely reducing the inoculum and the potential for disease.

Hot-Water Treatment of Propagative Organs

- Certain seeds, bulbs, and nursery stock are infected.
- Pathogen may be present inside seed coats, bulb scales, and so on, or which may be present in external surfaces or wounds.
- Hot-water treatment is used to kill any pathogens.
- Seed treatment with hot water was for many years the only means of control in some diseases.
- Loose smut of cereals, in which the fungus overwinters as mycelium inside the seed where it could not be reached by chemicals.

- Treatment of bulbs and nursery stock with hot water frees them from nematode such as *Ditylenchus dipsaci* in bulbs of various ornamentals and *Radolpholus similis* in citrus rootstockss within them.
- Temperature of the hot water used and the duration of the treatment vary with the different host–pathogen combinations.
- Thus, in the loose smut of wheat the seed is kept in hot water at 52°C for 11 minutes, whereas bulbs treated for *D. dipsaci* are kept at 43°C for 3 hours.
- Treated fruit had less soil, dust, and fungal spores at its surface while many of its natural openings in the epidermis were partially or entirely sealed.

Hot-Air Treatment of Storage Organs

- Treatment of certain storage organs with warm air (curing) removes excess moisture from their surfaces.
- It hastens the healing of wounds.
- Thus, preventing their infection by certain weak pathogens.
- For example, keeping sweet potatoes at 28 to 32°C for 2 weeks helps the wounds to heal and prevents infection by Rhizopus and by soft-rotting bacteria.
- Hot-air curing of harvested ears of corn, tobacco leaves, and so on removes most moisture from them and protects them from attack by fungal and bacterial saprophytes.
- Dry heat treatment of barley seed at 72°C for 7 to 10 days eliminates the leaf streak- and black chaff-causing bacterium *Xanthomonas campestris pv. translucens* from the seed with negligible reduction of seed germination.

Control by Eliminating Certain Light Wavelengths

- Alternaria, Botrytis, and Stemphylium are examples of plant pathogenic fungi that sporulate only when they receive light in the ultraviolet range (below 360 nm).
- It is possible to control diseases on greenhouse vegetables caused by several species of these fungi.
- It is done by covering or constructing the greenhouse with a special ultraviolet (UV)-absorbing vinyl film that blocks the transmission of light wavelengths below 390 nanometers.

Drying Stored Grains and Fruit

- All grains, legumes, and nuts carry with them a variety of plant pathogen.
- They can cause decay of these organs in the presence of sufficient moisture.
- Such decay, however, can be avoided
    - if seeds and nuts are harvested when properly mature.
    - Then are allowed to dry in the air.
    - or are treated with heated air until the moisture content is reduced sufficiently (to about 12% moisture) before storage.
- Subsequently, they are stored under conditions of ventilation.
- Fleshy fruits, such as peaches and strawberries, should be harvested later in the day, after the dew is gone, to ensure that the fruit does not carry surface moisture with it during storage and transit, which could result in decay of the fruit by fungi and bacteria.
- Grapes, plums, dates, and figs can be dried in the sun or through warm air treatment to produce raisins, prunes, and dried dates and figs, respectively, that are generally unaffected by fungi and bacteria as long as they are kept dry.

Disease Control by Refrigeration

- Refrigeration is probably the most widely used and the most effective method of controlling postharvest diseases of fleshy plant products.
- Low temperatures at or slightly above the freezing point do not kill any of the pathogens that may be on or in the plant tissues.
- They do inhibit or greatly retard the growth and activities of all such pathogens.
- Thereby reducing the spread of existing infections and the initiation of new ones.
- Most perishable fruits and vegetables should be refrigerated as soon as possible after harvest, transported in refrigerated vehicles, and kept refrigerated until they are used by the consumer.
- The magnitude of disease control through refrigeration and its value to growers and consumers is immense.

Disease Control by Radiation

- Various types of electromagnetic radiation, such as UV light, X rays, and γ rays, as well as particulate radiation, such as α particles and β particles are available.

- They are used for their ability to control postharvest diseases of fruits and vegetables.
- They kill the pathogens present on them.
- Some preliminary results are obtained in experimental studies using γ rays to control postharvest infections of peaches, strawberries, and tomatoes by some of their fungal pathogens.

Trench Barriers against Root-transmitted Tree Diseases

- Several vascular wilt and other diseases of trees are transmitted from tree to tree through contact or natural grafts between roots of adjacent trees.
- Several types of water permeable and nonpermeable materials are available and seem to be effective in blocking root contact.
- Water-permeable materials seem to be superior.
- However, the cost is high for either type of material and only highvalue landscape trees may justify such treatment.

3

Important Cultural Practices and Their Role in Desease Management

Cultural Management

Definition

- Manipulation of cultural practices at an appropriate time for reducing or avoiding disease damage to crops.
- Cultural practices make the environment less favorable for the plant pathogen and or more favourable for its bio control agents.

Types of Cultural Practices

Katan (1996) has divided cultural practices into three categories

1. Practices which are usually applied for agricultural purposes not related to crop protection, such as fertilization and irrigation. They may or may not have a positive or a negative side-effect on disease incidence.
2. Practices which are used solely or mainly for disease control, such as sanitation for the eradication of infected plant residues, and flooding.
3. Practices which are used for both agricultural purposes and disease control, such as crop rotation, grafting, and composting.

Effect of Cultural Practices on Disease Control

Direct effects

- By reducing the amount of inoculum and reducing the spread of an established pathogen.

Indirect effects

- By increasing the population of microbial agents in the soil which will be helpful in controlling the pathogens.

List of Cultural Practices

1. Soil solarization
2. Deep summer ploughing
3. Organic and inorganic amendments

4. Fallowing
5. Crop rotation
6. Green manure crops
7. Irrigation practices
8. Roughing
9. Strip farming
10. Trap and decay crops
11. Burning crop residue
12. Fertilizers usage
13. Time of sowing
14. Sanitation

Deep Summer Ploughing

- Ploughing the land during hot summer periods in early monsoon period.

Advantages

- Exposing of resting spores to the sun the sensitive pathogens die.
- Eg: Dormant structures of soil borne fungi;
  - **Dormant spores** (Conidia; Peach leaf curl pathogen. *Taphrina deformans*).
  - **Chlamydospores** (Wilt pathogen, Fusarium sp.)
  - **Oospores** (Downy mildew fungi)
  - **Perithecia** (Apple scab pathogen, *Venturia inaequalis*)
  - **Thickened hypha, sclerotia** (Cottony rot fungus, *Sclerotinia sclerotiorum*)
  - **Microsclerotia** (*Verticillium*)
  - **Rhizomorphs** (*Armillaria mellea*)

Fallow land

- Keeping of land vacant without any crop is known as "fallowing"
- During fallowing debris and plant pathogen inoculum are destroyed.
- Eg: Panama wilt of banana controlled by 6 years fallow.

Crop rotation

Definition

- Managing plant diseases by using another plant.
- Growing of different crops in a specific piece of land in a year after year.
- On a certain land, repeated cultivation of different crops or fallow land in certain sequence.

Purpose

- To manage the soil borne pathogens.
- To manage the obligate pathogens, where pathogen cannot survive away from the suitable host.
- To augment the biocontrol agents in the soil.
- Satisfactory control through crop rotation is possible with pathogens that are soil invaders.
- Soil invaders survive only on living plants or only as long as the host residue persists as a substrate for their saprophytic existence.
- When the pathogen is a soil inhabitant i.e., produce long lived spores or can live as a saprophyte for more than 5 or 6 years, crop rotation becomes less effective or impractical.

Soil inhabitants Vs Soil invaders

Soil inhabitants

- Unspecialized parasites with a wide host range.
- Able to survive indefinitely in the soil as saprophytes.
- Include obligate saprophytes and facultative parasites. They are exo-pathogens.
- Soil and plant debris serve as media for their saprophytic survival.
- They have high competitive saprophy tic survival ability.
- Species of Pythium, Rhizoctonia, Sclerotium, etc survive as soil inhabitant in the absence of the host.

Soil invaders/Root inhabiting fungi

- Specialized parasites that survive in soils in close association with their hosts.
- Include facultative saprophytes which are endo-pathogens (poot infecting fingi).
- The active saprophytic phase remains as long as the host tissue in which they were living as parasites is not completely decomposed.
- They have low competitive saprophytic survival ability.
- Most plant pathogenic fungi (Verticillium) and bacteria are soil invaders.

Examples of Diseases Control by Crop Rotation

- Potato cyst nematode:Mashua (*Tropaeolum tuberosum*), Tarwi (*Lupinus mutabilis*) these plants contains nematicidal properties.
- Soil borne pathogens: such as *Macrophomina* can controlled by growing smoothering crops.
- Fusarium wilt in redgram by sorghum.

Irrigation practices

Definition

- Artificial application of water to the soil is called "Irrigation"

Factor effecting the disease incidence

- Method of irrigation
- Irrigation frequency
- Number of irrigations

Examples

- **Flood irrigation**
- Due to this the spores spread from one to other field in citrus orchards and in paddy.
- **Sprinkler irrigation**
- Favours foliar diseases because of high humidity.
- Spread the foliar diseases like colletotrichum.
- **Double ring method**
- Reduces gummosis disease in orchards.
- Reduce the certain diseases like root rot caused by Rhizoctonia Bataticola.
- Drainage can reduce the inoculum of propagules in soil.

Irrigation and Drainage Water

- Sail-borne pathogens may be carried from one field to another field through irrigation or drainage water.
- Hence care should be taken not to irrigate a healthy field using drainage water from an infected field.
- Ex. Bacterial blight of rice (*Xanthomonas campestris pv. oryzae*) spreads mostly through irrigation and drainage water.

Roughing

- Removal and destruction of alternate hosts and collateral host, diseased plants, or rogues.
- It can prevent further spread of disease.
- Labour intensive.
- Practicable only in small plantings, or where labour is cheap or if the crop is very valuable.

- Some pathogens require two alternate hosts to complete their full life cycles.
  - *Puccinia graminis tritici* requires wheat and barberry.
  - *Cronartium ribicola* requires pine and currant (Ribes).
  - *Gymnosporangium juniperi-virginianae* requires cedar and apple.
- In these cases, eradication of the wild or economically less important alternate host interrupts the life cycle of the pathogen and leads to control of the disease.

Roguing Infected Plants

- Infected plants should be rogued out to reduce the inoculum in the field.
- Smutted whip should be removed to reduce sugarcane smut incidence.
- Virus-infected plants should be removed to reduce the virus inoculum.
- Yellow mosaic of greengram can be reduced by roguing infected plants.

Pruning of Infected Twigs and Branches

- Removal of infected twigs may reduce disease incidence.
- Pruning of citrus canker affected branches reduces the disease incidence.

Removal of Affected Portion

- Surgery is carried out to remove affected woody portion to control citrus gummosis caused by *Phytophthora citrophthora*.

Burning Crop Residue:

- Crop residue is burnt to avoid carry-over of inoculum in case of rice sheath blight (*Rhizoctonia solani*) and rice stem rot (*Sclerotium oryzae*).

Different between Collateral host and Alternate host

Collateral host / Alternative host

- Wild host of same family.
- For off season surviving of pathogen.
- To bridge the gap between two crop seasons.
- Eg: Paddy blast-*Panicum repens*.

Alternate host

- Wild host of other families.
- For completion of the life cycle of pathogen.
- Host compulsory required.
- Eg; Wheat black stem rust , Barbary leaf.

Trap or Decoy Crops

Trap Crops

- Trap crops of susceptible plants are grown on land known to contain pathogens.
- They become infected and are then destroyed before the pathogens' life cycles are complete, thus reducing the amount of inoculum in the area.
- Tagetes is a trap crop for many nematodes.
- It can be grown to reduce nematode infection.
- Fodder sorghum can be raised to reduce downy mildew of sorghum.

Decoy Crops

- Decoy crops stimulate the hatching or germination of pathogen eggs or propagules, but the pathogens are unable to establish and infection of the decoy host and die, again reducing the amount of available inoculum.
- Eg; Sunhemp.

Fertilizer Usage

- Fertilizer usage may have some bearing on development of certain diseases.
- It differs with each crop and each disease.
- In general, nitrogen out of balance with other nutrients enhances foliage disease development and predisposes some plants to other diseases.
- Potash, on the other hand, helps reduce disease development when it is in balance with other elements

Nitrogenous Fertilizers

a) Dosage

- Increased application of nitrogen increases many diseases.
- Rice blast becomes severe in nitrogen applied fields (18).

b) Method of application

- Split application reduces blast and bacterial blight of rice.
- Contrarily late application of nitrogen increases wheat leaf blotch (Septoria nodorum) and powdery mildew (Erysiphe graminis)

c) Type of N fertilizer

- Some diseases are favoured by ammoniacal form of nitrogen while others are favoured by nitrate form of nitrogen.
- Wilts (Fusarium spp.) and roots (Rhizoctonia spp.) are favoured by ammoniacal nitrogen.

- Verticillium wilts and root rots due to Pythium spp. are favoured by nitrate nitrogen.
- In rice, blast disease is favoured by ammoniacal nitrogen while brown spot (*Helminthosporium orzae*) is favoured by nitrate nitrogen.
- In maize leaf spot caused by *Helminthosporium turcicum* is favoured by ammonical nitrogen while stalk rot (*Diplodia maydis*) is favoured by nitrate nitrogen.
- In wheat, sharp eyespot (*Rhizoctonia solani*) is favoured by ammonical nitrogen while stem rust (*Puccinia graminis tritici*) is favoured by nitrate nitrogen.
- In potato, wilt caused by (*Verticillium albo-atrum* and scab caused by *Streptomyces scabies* are favoured by nitrate nitrogen while ammoniacal nitrogen increases black scurf (*R.solani*).

Phosphorous

- Repeated application of phosphatic fertilizers delays the onset and lessens the severity of take-all disease of barley caused by *Gacumannomyces graminis*).

Potassium

- Potassium application reduces the disease incidence in many cases probably by increasing phenolics synthesis in plants.

Calclum

- Calcium application reduces bean root rot caused by *Rhizoctonia solani* probably by altering pectin metabolism of the host .

Molybdenum

- Application of molybdenum reduces infection of potato tubers by *Phytophthora infestans.*
- Diminishes incidence of Ascochyta blight on beans and peas.

Manganese

- Manganese reduces late blight of potato.

Iron

- Ferric chloride controls rice brown spot.

Silicon

- Silicon application reduces rice blast caused by *Pyricularia oryzae.*

Time of Seeding

- Time of seeding has an important bearing on disease prevention in many cases.
- Delayed planting of wheat will help escape the chances of wheat streak mosaic virus.
- Early spring planting of cotton may effectively help escape cotton root rot.

Seeds

- Seeds carry many fungi, bacteria and viruses.
- May introduce these pathogens into the field.
- Hence, the infected seeds should not be used for sowing.
- Loose smut wheat (*Ustilago tritici*), black arm of cotton (*Xanthomonas campestris pv malvacearum)* and leaf crinkle of blackgram (virus) are introduced into the field through seed.

Vegetative Propagating Material

- Banana suckers may carry the bunchy top virus.
- Hence suckers from diseased areas should not be used.
- Sugarcane setts may carry red rot pathogen (*Colletotrichum falcatum*).
- Hence sugarcane setts from diseased areas should not be used for planting.

Seed Rate and Spacing

- Closer spacing favours many air-borne diseases because of high humidity in the crop canopy.
- Tikka leaf spot of groundnut (*Cercospora arachidis*) and blister blight of tea (*Exobasidium vexans*) are more in dense canopy.
- Damping off caused by Pythium and Phytophthora increases in crops when higher seed rate was used.
- Crowded stands may reduce some systemic diseases.
- Cotton wilt caused by *Verticillium albo-atrum* will be less in closely planted crop if the fungal inoculum is less in the soil.
- Similarly, closer spacing of rice reduces rice tungro infection particularly when vector population is less.

Sanitation

1. Sanitation consists of all activities aimed at eliminating or reducing the amount of inoculum present in a plant, a field, or a warehouse.
2. It preventing the spread of the pathogen to other healthy plants and plant products.

Disease Management by Cultural Practices

- Every operation of cultural practices may affect the disease incidence.
- Hence by proper cultural practices many diseases can be controlled.
- Some of the examples are given in this lecture.

1. Seeds

- Seeds carry many fungi, bacteria and viruses.
- May introduce these pathogens into the field.
- Hence, the infected seeds should not be used for sowing
- Loose smut wheat (*Ustilago tritici*), black arm of cotton (*Xanthomonas campestris pv malvacearum)* and leaf crinkle of blackgram (virus) are introduced into the field through seed.

2. Vegetative Propagating Material

- Banana suckers may carry the bunchy top virus.
- Hence suckers from diseased areas should not be used.
- Sugarcane setts may carry red rot pathogen (*Colletotrichum falcatum*).
- Hence sugarcane setts from diseased areas should not be used for planting.

3. Irrigation and Drainage Water

- Sail-borne pathogens may be carried from one field to another field through irrigation or drainage water.
- Hence care should be taken not to irrigate a healthy field using drainage water from an infected field.
- Ex. Bacterial blight of rice (*Xanthomonas campestris pv. oryzae*) spreads mostly through irrigation and drainage water.

4. Plant Debris

- Some of the pathogens survive in plant debris.
- Hence, plant debris should be burnt or removed.
- Eg. Stem rot of rice (*Sclerotium oryzae*) and black arm of cotton (*X campestris pv. malvacearum*).

5. Equipment and Men

1. Pruning, implements, intercultivators, etc., may carry some pathogens.
2. Workers also may carry the virus from infected field to the healthy field.
3. Hence proper sanitation is needed.
4. Eg, Tobacco mosaic virus in tobacco field.

6. Weeds

- Weeds serve as alternate hosts for many pathogens.
- Hence clean cultivation is essential for control of diseases.
- Eg, Rice tungro virus, Rice blast, etc.

7. Volunteer Plants

- Self-sown volunteer plants may carry infection which may serve as source of infection for the succeeding crop
- Volunteer plants should be destroyed.
- Eg, cotton black arm.

8. Ratoons

- Ratoon crops normally carry many pathogens.
- Ratoon crops should be avoided as far as possible.
- Eg, Sugarcane grassy stunt.

13. Solar Heating

- When the soil is covered with polyethylene sheets during hot seasons.
- Soil temperature increased and eliminated *Verticillium dahliae* and *Fusarium oxysporum f. sp. lycopersici* in tomato field.
- High temperature favours growth of antagonistic fungi also.

14. Flooding the Field

- *Fusarium oxysporum f. sp. cubense* does not survive well in flooded field.
- Hence Fusarium wilt of banana can be controlled by flooding the field for long time.
- Tobacco black shank (*Phytophthora parasitica var. nicotianae*) and Cotton wilt (*Verticillium dahliae*) can also be controlled by flooding the field.

15. Fallowing

1. *F. oxysporum* specific strains can be reduced as they do not survive in soil for long time.
2. But fallowing will not reduce *Rhizoctonia solani* infection.

16. Date of Sowing

- By adjusting the time of sowing some of the diseases can be controlled.
- Eg, early sowing of pearl millet will reduce the ergot disease.
- Avoiding wet seasons for planting of rice will eliminate rice blast.

17. Avoidance of Monoculture

- Continuous cultivation of rice crop may result in the appearance of new pathotypes of *Pyricularia oryzae* the rice blast pathogen.

18. Multiple Cropping

- Taller crops can be sown to protect a crop of lesser height from virus vectors.
- Insects may land at the taller crops and the dwarf crop may escape from virus diseases transmitted by those insects.

19. Intercropping

- All the crops should not have any common pathogen.
- Intercrops should be properly chosen.
- For example, *Rhizoctonia solani* has wide host range.
- Hence common hosts of the pathogen should not be grown as intercrops.

20. Trap Crop

- Tagetes is a trap crop for many nematodes.
- It can be grown to reduce nematode infection.
- Fodder sorghum can be raised to reduce downy mildew of sorghum.

21. Tillage

- Ploughing of rice residue to a depth of 25-30 cm renders a high percentage of inoculum of *Sclerotium oryzae* .
- It is unavailable for infection of rice subsequently sown with flooding.

22. Soll Amendment

- Lucerne meal and barley straw reduce root rot of cotton caused by *Macrophomina phaseolina.*
- Potato black scurf (*Rhizoctonia solani*) is less in soil amended with wheat straw.
- Pea root not (*Aphanomyces euteiches*) is reduced when cruciferous plant residue is incorporated in soil.

23. Seed Rate and Spacing

- Closer spacing favours many air-borne diseases because of high humidity in the crop canopy.
- Tikka leaf spot of groundnut (*Cercospora arachidis*) and blister blight of tea (*Exobasidium vexans*) are more in dense canopy.
- Damping off caused by Pythium and Phytophthora increases in crops when higher seed rate was used.

- Crowded stands may reduce some systemic diseases.
- Cotton wilt caused by *Verticillium albo-atrum* will be less in closely planted crop if the fungal inoculum is less in the soil.
- Similarly, closer spacing of rice reduces rice tungro infection particularly when vector population is less.

24. Irrigation

- Potato scab (*Streptomyces scabies*) is prevented by maintaining soil water near field capacity during the tuber formation.
- High soil moisture content encourages antagonistic bacterial population.
- Irrigation controls *Macrophomina phaseolina* infection in potato and cotton.

4

Biological Control of Plant Diseases

Biological control

Definition

Garret (1970) defined biological control as any condition under which or practice whereby survival or activity of a pathogen is reduced through the agency of any other living organism (except man) with the result that there is a reduction in the incidence of the disease caused by that pathogen.

Baker and Cook (1974) stated that the biological control is the reduction of inoculum density or disease producing activities of a pathogen or parasite in its active or dormant stage, by one or more organisms, accomplished naturally or through manipulation of the environment, host, or antagonists, or by mass introduction of one or more antagonists.

Biocontrol can be simply defined as the application of living organism to control diseases.

Differences between disease bio-control and insect bio-control

Sl.No	Disease bio-control	Insect bio-control
1	Disease control is mainly achieved by antibiosis, competition and by hyperparasites.	Mainly by parasites and predators.
2	Antagonists are mainly passive and are not mobile. Contact of pathogen is accidental.	Parasites are active, mobile and seek their prey.
3	It is a mass effect. For a single species of pathogen a large number of antagonists / competitors available.	Single predator / parasite for single prey.
4	This method relies mainly on native organisms.	Introduction of parasites / predators from other countries are normally followed.
5	Pathogen free seeds and planting materials are widely used.	Pest free seeds are not used.

Mechanism of bio control

1. Competition
2. Hyperparasitism
3. Antibiosis

4. Antagonism
5. Exploitation
6. Induction of host resistance
7. Hypovirulence

Competition

1. All kinds of interplay between organisms in which one is favored at the expense of the other.
2. Defined as a more or less active demand in excess of the immediate supply of material or condition on the part of two or more organism.

Antibiosis

1. Condition in which one or more metabolites excreted by an organism have harmful effects on one or more other organism.
2. In such antagonistic relationship species A produces a chemical substance that is harm or inimical to species B without species A deriving any direct benefit.
3. However, the species A may have an indirect benefit in having a better competitive ability, thereby getting an advantage over species B for substrate colonization.

Mycoparasitism

1. An act where one fungus parasitizes the other one.
2. Mycoparasitism includes different kinds of interaction, viz. coiling of hyphae, penetration, production of haustoria and lysis of hyphae.
3. Mycoparasitism is of common occurrence and examples can be found among all groups of fungi from chytrids to the higher basidiomycetes.

Antagonism

1. Any association of two or more species, at least one of the interacting species is harmed due to the activities of one or more of the rest.

Exploitation

1. When species A inflicts harm by the direct use of species B for its own benefit, it is exploitation (parasitism and predation).
2. The two terms parasitism and predation have basically same effects, i.e., destruction of the host or prey.
3. In parasitism, some sort of etiological relationship between the parasite and the host is established and the host is not rapidly eliminated.
4. A predator physically eliminates its prey by direct feeding on it without establishing any etiological relationship.

Hypovirulence

1. Phenomenon of reduced virulence of a pathogen strain than normal one developed as a result of its infection by ds RNA.

Induction of host resistance

1. Plants respond to a variety of chemical stimuli produced by soil- and plant-associated microbes.
2. Such stimuli can either induce or condition plant host defenses through biochemical changes that enhance resistance against subsequent infection by a variety of pathogens

Some examples of mechanism of biocontrol

Sl.No	Mechanism	Examples
1	Hyperparasitism / Mycoparasitism	• *Trichoderma harzianum* paraitize mycelia of *Rhizoctonia and Sclerotium*
2	Antibiosis	• HCN produced by *Pseudomonas fluorescens* • Bacilysin produced by *Bacillus subtilis* • Trichodermin produced by *Trichoderma*
3	Competition	• Exudates/leachates consumption • Siderophore scavenging • Physical niche occupation
4	Induction of host resistance	• ISR mediated resistance in tobacco, Aridiopsis • Rhizobacteria applied to seeds or roots induce systemic resistance response expressed against pathogens infecting aerial tissues.
5	Hypovirulence	• Hypovirulence of fungal pathogens by mycovirus.

Role of Biocontrol

1. Disease control
2. Plant growth promoter
3. Stimulation of plant resistance and plant defense mechanism
4. Transgenic plants
5. Bioremediation

Advantages of Biocontrol

1. Perform preventive control
2. Low persistence and residual effect
3. Mostly biodegradable and self-perpetuating
4. Delayed knock down effect
5. Less resistance prone

6. Less pest resurgence
7. Less harmful on beneficial microorganisms
8. Mostly target specific
9. Waiting period almost nil
10. Costlier but reduced number of applications required

Limitations

1. Lack of faith and inconsistent field performance
2. Poor quality and shelf life
3. Competition with chemical pesticides
4. High budget of production and lesser agribusiness
5. Registration process is expensive and also time- consuming
6. Lack of knowledge, confidence and awareness in end user
7. Inclination towards use of chemical pesticides.
8. Non availability of popular literature
9. Mass production of biotic agents limited techniques for indigenous and exotic use
10. Laboratory evaluation of biotic agents and their field efficiency.
11. Transfer of technology requires more attention

Microbial Consortia

- When two or more microbial groups living symbiotically.

Advantages of Consortium

- Robustness
- Efficiency
- Modularity

Essential for success of Microbial Consortia

- Compatibility of microbial components
- Synergism of microbial components

Mode of action of Microbial Consortia

- Competition
- Mycoparasitism
- Antibiosis
- Antagonism

- Additive or synergistic effect
- Induction of host

Various applications of Microbial Consortia

- Biocontrol agents
- Biofertilizers
- Rhizosphere bioremediation of pesticides
- Assisting mammals in food digestion.
- Reclamation of soils
- Effective degradation of organic wastes

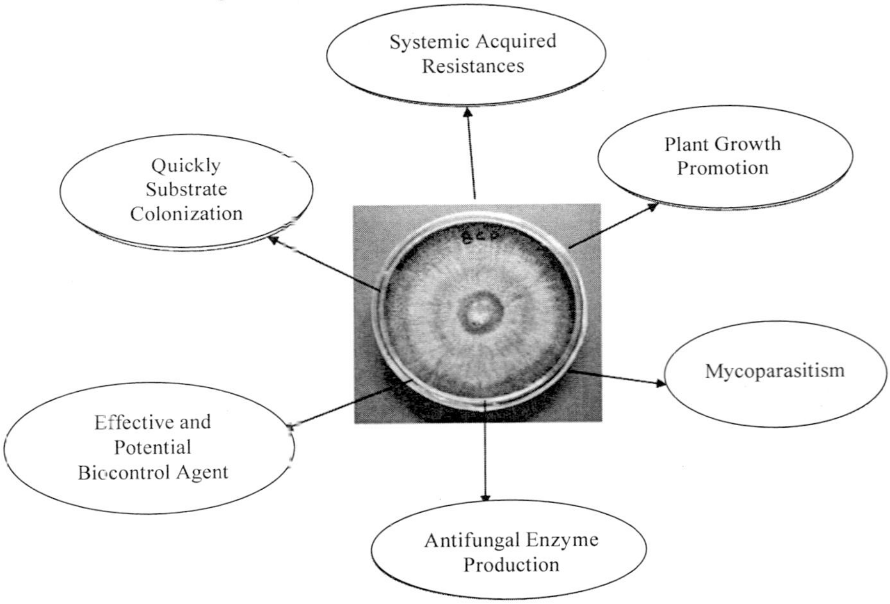

Figure 1. Biocontrol activities exhibited by *Trichoerma.*

Important biocontrol agents

1. Trichoderma spp
2. Pseudomonas
3. Bacillus
4. Pecilomyces

Factors affecting bioefficacy of Trichoderma spp

- Trichoderma spp. are widely distributed all over the world.
- It occur in all soils.

- *Trichoderma harzianum* and *Trichoderma viride* are the two important Trichoderma species which have been developed as biocontrol agents.
- Conidia of Trichoderma do not survive well in soil but chlamydospores survive for a long time.
- Fungistasis suppresses the growth of Trichoderma in soil.
- If Trichoderma are to be utilized for biological control, fungistasis in soil should be annulled.
- Organic amendments may annul fungistasis.
- Organic amendments do not support growth of Trichoderma because these organic amendments support other saprophytes which suppress Trichoderma.
- Trichoderma cannot compete with other saprophytes and it is only a secondary invader.
- Mature compost supports Trichoderma well.
- Trichoderma proliferates when added as young mycelia in intimate contact with a food base, but not conidia with or without bran.
- Young hyphae are less semitive to fungistasis.
- The thorough colonization of the food base by the mycelia enables Trichoderma to grow well through the soil.
- Thorough primary colonization of a substrate by Trichoderma before addition to soil is necessary.
- Alginate pellets containing Trichoderma established in the bran medium also stimulate the establishment of Trichoderma in soil.
- If soil is fumigated with carbon disulfide or formaldehyde or methyl bromide Trichoderma multiplies well in the soil.
- Sodium azide treatment also increases multiplication of Trichoderma in soil.
- These chemicals are non-inhibitory to Trichoderma but inhibit other saprophytes.
- Covering the soil with polythene sheets enhances soil temperature (solarization) and this sub-lethal beating of soil encourages Trichoderma.
- Thiram, captan, PCNB, chlorothalonil and chloroneb are not inhibitory to Trichoderma but suppress other saprophytes.
- Benomyl, carbendazim, thiophanate methyl and thiabendazole are inhibitory to Trichoderma.
- Trichoderma does not survive in the rhizosphere.

- Pseudomonas spp. compete with Trichoderma for iron in the rhizosphere and for want of iron the fungus does not survive in the rhizosphere.
- Trichoderma is aerobic and requires oxygen for its growth.
- The fungus does not survive in anaerobic soils (paddy fields).

Mode of Action of Trichoderma

- Trichoderma produces three antibiotics, viz. trichodermin, liotoxin and viridin in culture.
- But they are not produced in appreciable quantities in soil.
- Trichoderma invades hyphae of Pythium and Rhizoctonia and produces haustoria.
- It produces ß-(1-3)-glucanase and chitinase and degrades
  - Glucans in the cell walls of Pythium spp.
  - Chitin and glucans in the walls of Rhizoctonia.
- Thus these enzymes lyse the hyphae of the pathogens.

Limitations in the use of Trichoderma as Biocontrol Agents

- Trichoderma does not survive in rhizosphere as well as in unamended soil.
- The fungal inoculum also should be in young mycelial stage and old cultures with spores will be inactive. Hence storage of active cultures is difficult.
- Steaming or fumigating the soil to encourage Trichoderma will be expensive. Polythene mulching will also be expensive.
- Trichoderma doesn't survive in anaerobic soil. Hence, it is not useful against rice diseases.
- As seed treatment Trichoderma effectively controls soilborne diseases occurring in early stages of crop growth.
- However, many seed treatment fungicides inhibit Trichoderma.

Siderophores producing Bacteria

Siderophores

- *Pseudomonas fluorescens* and *P. putida* produce siderophores.
- These two bacteria control soft rot of potato caused by *Erwinia carotovora*.
- Siderophores are low molecular-weight ferric iron transport agents.
- The function of siderophores is to supply iron to the bacterial cell.
- Siderophores selectively complex with iron with very high affinity.

- So make it iron unavailable to other microorganisms including pathogens.
- Siderophores of Pseudomonas efficiently sequester ferric iron in potato root zone making it unavailable to other rhizoplane microorgnisms, including plant pathogens.
- Hence the populations of the pathogen are reduced and the roots appear healthier.
- Siderophore of Pseudomonas sp. is fluorescent and it is called Pseudobactin.
- Pseudobactin has been found to control take-all disease of wheat and barley caused by *Gaeumannomyces var. tritici* and flax wilt disease caused by *Fusarium oxysporum f. sp. lini*.
- Inoculation of flax seeds with a suspension of Pseudomonas sp. or pseudobactin controls Fusarium wilt disease.
- Addition of pseudobactin to the soil also controls wilt diseases.
- The pseudobactin converts pathogen-conducive soils into pathogen suppressive soils.
- It suggests that suppressiveness of the soil is mainly due to the siderophores.

Biological control against Soil-borne diseases

Some of the term often used in biological control

Soll Fungistasis

- When the fungal resting structures (sclerotia, chlamydospores, oospores, conidia, etc.) are added to the soil they do not germinate and grow in the soil.
- This condition is called soil fungistasis.
- Soil fungistasis is widespread natural soil toxicity, which inhibits propagule germination.
- Fungistasis is controlled by a complex balance of stimulators and inhibitors in soil micro environments.
- Fungistasis in soil can be annulled (removed) by addition of nutrients to soil (crop residues, root-exudates and various amendments).

Suppressive Soils

- Soil-borne pathogens do not survive and cause diseases in some soils.
- When the spores germinate after the annulment of fungistasis, the hypha

gets lysed or again dormant spores are formed.

- This kind of soils are called suppressive soils.
- Suppressiveness has been attributed due to several factors such as:
    - Antagonistic effects of associated microfloras.
    - Faster multiplication of bacteria in suppressive soils.
    - Elevated fungistasis to fungal propagules.
    - Microbial competition for food.
    - Ethylene production in soil.
    - Unidentified biological factors that affect formation and persistence of resulting spores.
- *Pseudomonas. Chromobacterium* and *Arthrobacter* are the most important bacteria which are responsible for suppressive soil.

Hyperparasites

- Fungal hyperparasites are also called 'mycoparasites.
- Hyperparasites are fungi which parasitize other fungi attacking both hyphae and sporulating structures.
- *Darluca, Tuberculina, Verticillium* und *Trichoderma* are the important hyperparasites which parasitize;
    - Uredo and teleutosori of numerous rust species.
    - Conidia of powdery mildews.
    - Hyphae of Rhizoctonia species.

Contagious Hypovirulence

- Some of the viruses which are called 'mycoviruses' attack fungi and cause reduction in their virulence.
- This phenomenon is called contagious hypovirulence.

Introduction of Foreign Antagonists

- When the soil-borne *Trichoderma viride* is applied on grapevine, it colonizes dead plant remains such as flowers which would otherwise provide a food base for Borrytis infection. Boytis rot of grapevine in controlled by this method.
- *Botrytis cinerea,* eyespot of apples is controlled by spraying conidia of *Trichoderma pseudokoningi.*
- Tree wound application of *T. viride* is known to control silver leaf disease of plum caused by *Stereum purpureum.*

5

Chemical Methods of Disease Management

Use of Chemical Pesticides

Objective

- To protect plant surfaces from infection.
- To eradicate a pathogen that has already infected a plant.
- To eradicate or greatly reduce the inoculum before it comes in contact with the plant.

Methods of Applications of Chemicals for Diseases Management

1. Soil application
2. Seed treatment
3. Foliar or vegetative application

1. Soil application

- Purpose is to reduce or eradicate the inoculums density of the pathogens present in *soil* or *soil invaders* or as *soil inhabitants.*
- Chemical soil treatment can be done in any of the following ways.

Soil Drenching

- Fungicidal suspension or solution is applied to soil surface in quantities enough to wet 10-15 cm depth of the soil before or after planting.

Broadcasting

- Non-volatile fungicides (dusts, powders, granules, etc.) are mixed with soil or fertilizers and spread on the soil surface.
- Light ploughing or harrowing is then done to mix the chemical in sufficient depth.

Furrow Application

- Fungicides are applied either as dusts or with water to the furrow at the planting time.
- Method is possible in crops planted in furrows/rows such as potato and sugarcane.

- Good for control of diseases that occur at the base of the plants.
- Requires much less quantity of the chemicals per hectare than the broadcast method.

Fumigation

- Usually used to control plant parasitic nematodes.
- Chemicals used are generally volatile and on coming in contact with soil moisture release gases which diffuse in the soil and kill the nematodes.
- Special equipments are required for application since diffusion is restricted to certain distance.
- Chemicals are injected into soil at regular distance (12-15 inch) all over the field. The depth of application is kept at 6-9 inches.

Soil Treatment with Chemicals

- Soil to be planted with vegetables, strawberries, ornamentals, trees, or other high-value crops, such as tobacco.
- It is frequently treated with chemicals for control primarily of nematodes, but occasionally also of soilborne fungi,
- Certain fungicides are applied to the soil as dusts, liquid drenches, or granules to control damping-off, seedling blights, crown and root rots, and other diseases.
- Fungicides used for soil treatments include metalaxyl, diazoben, pentachloronitrobenzene (PCNB), captan, and chloroneb.
- In fields where irrigation is possible, the fungicide is sometimes applied with the irrigation water, particularly in sprinkler irrigation.
- Most soil treatments, however, are aimed at controlling nematodes, and the materials used are volatile gases or produce volatile gases (fumigants) that penetrate the soil throughout (fumigate).
- Some nematicides, however, are not volatile but, instead, dissolve in soil water and are then distributed through the soil.

Fumigation

- Fumigation is done through the use of chemicals usually called fumigants.
- Chemicals commonly used for fumigtion are chloropicrin, methyl bromide, dazomet, and metam sodium.
- They either volatilize as they are applied to the soil or decompose into gases in the soil.

- These materials are general purpose pre plant fumigants.
- They are effective against a wide range of soil microorganisms, including nematodes, many fungi, insects, certain bacteria, and weeds.
- Contact nematicides, such as fensulfothion, carbofuran, ethoprop, and aldicarb, are of low volatility, are effective against nematodes and insects.
- They can be applied before and after planting of many crops that are tolerant to these chemicals.
- Nematicides used as soil fumigants are available as liquids under pressure, liquids, emulsifiable concentrates, and granules.
- These materials are applied to the soil either by spreading the chemical evenly over the entire field (broadcast) or by applying it only to the rows to be planted with the crop (row treatment).
- Nematicide is sealed in the soil instantly by a smoothing and firming drag or can be mixed into the soil with disk harrows or rototillers.
- Highly volatile nematicides are covered immediately with polyethylene sheeting which should be left in place for at least 48 hours.
- When small areas are to be fumigated, the most convenient method is through injection of the chemical with a hand applicator under a tarp that has been placed over the area.
- The edges of the tarp are covered with soil prior to injection of the chemical.
- Applications may also be made by placement of small amounts of granules in holes or furrows six inches deep, 6 to 12 inches apart, which should be covered immediately with soil.
- Effectiveness of the fumigants is based on their diffusion in a gaseous state through the pores of the soil throughout the area in which nematode and other pest control is desired.
- Distance the vapors move is influenced by the size and continuity of soil pores, by
    - Soil temperature (Best range is between 10 and 20°C).
    - Soil moisture (Best at about 80% of field capacity).
    - Type of soil (More material is required for soils rich in colloidal or organic matter).
    - Properties of the chemical itself.
- Nematicides with low volatility, such as carbofuran, do not diffuse through the soil to any great extent and must be mixed with the soil

mechanically or by irrigation water or rainfall.

- Most nematicides can be applied in irrigation water when it is provided as trickle soaks or drenches, except for the highly volatile methyl bromide and chloropicrin.
- Only low-volatility nematicides can be applied through overhead sprinkler systems.
- Chemical nematode control in the field is generally obtained by preplant soil fumigation with one of the nematicides applied only before planting.
- All nematicides are extremely toxic to humans and animals and should be handled with great caution.

Disinfestation of Warehouses

- Stored products can be protected from becoming infected by pathogens left over in the warehouse from previous years by
  - First cleaning thoroughly the storage rooms.
  - By removing and burning the debris.
- The walls and floors are washed with bleach, a copper sulfate solution or some other sanitizing agent.
- Warehouses that can be closed airtight and in which the relative humidity can be kept at nearly 100% while the temperature is between 25 and 30°C.
- They can be fumigated effectively with chloropicrin (tear gas).
- In all cases the fumigants should be allowed to act for at least 24 hours before the warehouse doors are opened for aeration.

Control of Insect Vectors

- When the pathogen is introduced or disseminated by an insect vector, control of the vector is as important.
- Application of insecticides for the control of insect carriers of fungus spores and bacteria has been fairly successful.
- Controlling diseases by killing the insect vectors with insecticides after they have arrived at the crop has seldom proved adequate.
- Even with good insect control, enough insects survive for sufficiently long periods to spread the pathogen.
- Nevertheless, appreciable reduction in losses from certain such diseases has been obtained by controlling their insect vectors, and the practice of good insect control is always desirable.
- Success in reducing virus transmission by insects has been achieved by interfering with the ability of the aphid vector to acquire and to transmit the virus rather than by killing the insects.

- The interference is provided by spraying the plants several times each season with a fine-grade mineral oil. Such oil seems to have little effect on the probing and feeding behavior of the aphids and is not particularly toxic to the aphids, but it interferes with the transmission of nonpersistent, semipersistent, and even some persistent aphid-borne viruses. Control of aphid-borne viruses by oil sprays has been successful with some viruses (e.g., cucumber mosaic virus on cucumber and pepper and potato virus Y on pepper).

2. Seed treatment

- Seeds, tubers, bulbs and other propagating materials are often treated with fungicides.
- Purpose is to eradicate pathogen propagules present on and/or in them as well as for preventing their rot in the soil after planting.
- Chemicals can be applied on the seed as dusts, or as thick water suspension (slurry) mixed with the seed.
- On the basis of their mode of action of the seed treatment chemicals can be of three types:

Seed disinfestants

- Those which kill the pathogen present on the seed and do not remain active for long after the seed has been planted.

Seed disinfectants

- Those which act as eradicants and destroy the pathogen established in seed tissues.
- A number of systemic chemicals are being used for this purpose.

Seed protectants

- Seed are treated with protective fungicides.
- Pathogens present on the seeds are destroyed.
- Compounds remain on seed surface for sufficiently long times.
- It should also be ensured that their viability is not lowered or destroyed.

Seed biopriming

- Treating of seeds with biocontrol agents and then incubating under warm and moist conditions until just prior to emergence of radical is reffered as bioprimming.
- This technique has potential advantages over simple coating of seeds as it results in rapid and uniform seedling emergence.

- *Trichoderma* conidia germinate on the seed surface and form a layer around bioprimed seeds.
- Such seeds tolerate adverse various soil conditions better.
- Biopriming could also reduce the amount of biocontrol agents that is applied to the seed.

Spraying

- This is the most commonly adopted method.
- Spraying of fungicides is done on leaves, stems, flowers and fruits.
- Wettable powders are most commonly used for preparing sprays.
- Most common diluent or carrier is water.
- Dispersion of the spray is usually achieved by its passage under pressure through nozzles of sprayers.
- Spraying is of two types, namely, high volume and low volume.
- When sprays involve large quantities of liquid per unit area, they are termed high volume.
- Six hundred litres and above per hectare would be considered to be of the high volume category.
- With low volume sprays, it is usually possible to cover one hectare with about 100 litres or less.

Dusting & Paster or Paints

Dusting

- Dusts are applied to leaves, fruits, and stems of plants as an alternative to spraying.
- Dry powders are used for covering host surface.
- Dusting is practicable only in the calmest weather and high effectively is obtained if the dust is applied when the plant is wet with dew or rain.

Paster or Paints

- In orchards, the wounds created during pruning and training of trees often serve as entry points of the pathogens.
- To protect these wounds fungicidal pastes or paints are used as protective layer.
- These pastes are prepared with fungicides in a suitable carrier such as raw linseed oil lanolin, glycerine, etc.
- Residual effect of the pastes lasts long enough to permit natural healing of the cut surfaces.

6

Chemical Nature and Classification of Fungicides

Chemical Nature

- Fungicide has originated from two Latin words: viz. *fungus and caedo (to kill).*
- Fungicide mean a *chemical capable of killing fungi.*
- Chemicals inhibit growth or spore germination temporarily is called a *fungistat.* and the phenomenon as *fungistasis.*
- Bordeaux mixture may inhibit spore production without affecting vegetative growth. These are called *antisporulants* .
- There are other groups of chemicals which exhibit very poor or no antifungal activity *in vitro* condition but provide protection to the plants either by inhibiting the penetration of host surface by the fungi or by inducing the host defense system. The former type of chemicals are termed as *antipenetrants* and later as *antipathic agents.*
- Fungicide which is effective only if applied prior to fungal infection is called *protectants.*
- On the other hand, fungicide which is capable of eradicating a fungus after it has caused infection, and thereby curing the plant, is called *therapeutant.*

Characters of chemicals or fungicides

- High field performance.
- Easy availability of the active constituents.
- Absence of toxicity for the host, man and animals.
- High toxicity for the pathogen at low concentration.
- Retention of toxicity on dilution.
- Stability in storage.
- Slow or no loss of toxicity in storage.
- Good spreading quality on host surface.
- High tenacity on the host surface.

Advantages and disadvantages of chemicals or fungicides

Sl.no	Advantages	Disadvantages
1	Quick-acting	Potentially toxic to humans
2	Easy to use	Impact on non-targets
3	Cheap (relatively)	Potential negative environmental effects
4	Readily available	Unknown cumulative effects
5	Can save lives (vector-borne diseases)	Pest resistance

Classification of Chemicals

The chemicals can be broadly classified on the basis of their

- Mode of action against pathogen
- Type of pathogen.

According to mode of action

1. Protectants

- Protectants are prophylactic in their behaviour.
- They act outside the plant parts as a cover to check the invasion by the pathogen.
- They include thiram, captan, agallol, zineb, sulphur, ceresan and streptocyclin.

2. Eradicants

- Eradicants help to eradicate the dormant or active pathogen from the host completely.
- These chemicals can be used as protectant as well as eradicant.
- Examples are lime, sulphur, organo mercurials.

3. Therapeutants

- Therapeutants is an agent that inhibits the growth and development of a disease already entered in a plant, when applied appropriately.
- Usually the chemo-therapeutants are systemic in their action, i.e. they enter the plants and affect deep-seated infection.
- e.g. plantvax, vitavax, dithane M-45, streptocyclin and agromycin.

According to type of pathogen

1. Fungicides
2. Bactericides
3. Nematicides

Properties of Contact and Systemic Fungicides

Sl.No	Contact fungicides	Systemic fungicides
1	Multisite inhibitors	Single site inhibitors
2	Protective in function	Curative in function
3	Many metabolic systems of fungal pathogens affected	Many metabolic systems of fungal pathogens affected
4	Phytotoxicity is common	Phytotoxicity is rare
5	Target non-specific	Target specific
6	Pathogen resistance development rarely takes place	Pathogen resistance development commonly takes place
7	They are localized in action	They are systemic in action
8	They are confined to redistribution on plant surfaces	They are translocated inside the plant

7

Contact Fungicides

Contact Fungicide

Contact fungicide remain on the leaf surface after application and don not penetrate the tissues. These fungicides are usually used for the control of foliar diseases. A contact fungicide must be present on the leaf surface before the disease penetrate the tissues.

Inorganic Chemicals

1. Copper compounds
2. Inorganic sulfur compounds
3. Carbonate compounds
4. Phosphate and phosphonate compounds
5. Film-forming compounds

Organic Chemicals

1. Contact protective fungicides
2. Organic sulfur compounds: Dithiocarbamates.
3. Ethylenebisdithiocarbamates
4. Quinones
5. Aromatic compounds
6. Heterocyclic compounds

Inorganic Chemicals

Copper Compounds

Bordeaux mixture

1. Named after the Bordeaux region of France.
2. Developed in France and used against the downy mildew of grape.
3. First fungicide to be developed and still is the most widely used copper fungicide throughout the world.
4. Product of reaction of copper sulfate and calcium hydroxide (hydrated lime).

5. Controls many fungal (including oomycete) and bacterial leaf spots, blights, anthracnoses, downy mildews, and cankers.
6. Can cause burning of leaves or russeting of fruit such as apples when applied in cool, wet weather.
7. Phytotoxicity is reduced by increasing the ratio of hydrated lime to copper sulfate.
8. Copper is the only ingredient in the Bordeaux mixture that is toxic to pathogens.
9. Sometimes toxic to plants, whereas the role of lime is primarily that of a "safener."
10. Fixed coppers are used for control of the same diseases as Bordeaux.
11. Fixed coppers contain basic copper sulfate, sold as
12. Microcop and many other names, Copper oxy chlorides, Copper hydroxide, Copper oxides and Copper ammonium carbonate.

Inorganic Sulfur Compounds

- Element sulfur is oldest fungicide known.
- Used primarily to control powdery mildews, certain rusts, leaf blights, and fruit rots.
- Sulfur may cause injury in hot (temperatures above 30°C) & dry weather.
- Sulfur-sensitive plants such as tomato, melons, and grape.
- By boiling lime and sulfur together, lime–sulfur, selfboiled lime–sulfur, and dry lime–sulfur are produced.
- These, sold as Lime Sulfur, Orthorix, Sulforix, or Polysul.
- Used as sprays for dormant fruit trees to control blight or anthracnose, powdery mildew, apple scab, brown rot of stone fruits, and peach leaf curl, and are sometimes used for summer control of the same diseases.

Carbonate Compounds

1. Sodium bicarbonate as well as bicarbonate salts of ammonium, potassium, and lithium plus 1% superfine oil.
2. Fungicidal to the powdery mildew fungi of roses, black spot fungus of roses, to southern blight fungus *Sclerotium rolfsii*, and to the gray mold fungus *Botrytis cinerea.*

Phosphate and Phosphonate Compounds

1. Monopotassium phosphate (KH2PO4) or Dipotassium phosphate (K2HPO4)

2. Gave satisfactory control of the powdery mildew diseases of cucumber or grape plants with solutions of either .

Film-Forming Compounds

- Film-forming compounds.
- Antitranspirant polymers, mineral oils, surfactants, and kaolin-based particle.
- Applied on plant surfaces before inoculation with the pathogen reduce the number of infections significantly.
- They seem to reduce infections by altering the characteristics of the leaf surface.
- Interfering with the adhesion of the pathogen to the host and with the recognition of infection sites on the host.

Organic Chemicals

Organic Sulfur Compounds

Dithiocarbamates

- Organic sulfur compounds are one of the most important, versatile, and widely used group of modern fungicides.
- They are all derivatives of dithiocarbamic acid.
- Dithiocarbamates are toxic to fungi because they are metabolized to the isothiocyanate radical (—NKCKS).
- This radical inactivates the sulfhydryl groups (—SH) in amino acids and in enzymes within pathogen cells, thereby inhibiting the production and function of these compounds.
- They include thiram, ferbam, nabam, maneb, zineb, and mancozeb.
- Thiram consists of two molecules of dithiocarbamic acid joined together.
- It is used mostly for seed and bulb treatment for vegetables, flowers, and grasses ,rusts of lawns.
- Ferbam consists of three molecules of dithiocarbamic acid reacted to one atom of iron.
- Used to control foliage diseases, especially fruit on trees and ornammental plants.

Ethylenebisdithiocarbamates

- Another group of dithiocarbamic acid derivatives with different molecular configurations.
- Contains the fungicides maneb and zineb.

1. Maneb

- Contains manganese.
- Excellent, broad-spectrum fungicide.
- Controls foliage and fruit diseases of many vegetables,and some fruit.
- Maneb is sometimes mixed with zinc or with zinc ion and results in the formulations known as maneb zinc (sold as Manzate D).

2. Mancozeb

- Zinc ion maneb, called mancozeb.
- Addition of zinc reduces the phytotoxicity of maneb and improves its fungicidal properties.
- Secondary effect of the use of mancozeb is to supply Mn and Zn to deficient plants.
- Zineb is sold as Dithane Z-78.
- Excellent, safe, multipurpose foliar and soil fungicide for the control of leaf spots, blights, and fruit rots of vegetables, flowers, fruit trees, and shrubs

Quinones

- Occur naturally in many plants.
- Also, produced upon oxidation of plant phenolic compounds.
- Often show antimicrobial activity.
- Often considered to be associated with the innate resistance of plants to disease.
- Only two quinone compounds, chloranil and dichlone.

Aromatic Compounds

- Many rather unrelated compounds that have an aromatic (benzene) ring are toxic to microorganisms.
- Several have been developed into fungicides and are used commercially.
- Most seem to inhibit production of compounds that have —NH2 and —SH groups, namely amino acids and enzymes.
- Pentachloronitrobenzene, sold as PCNB Terraclor, Engage, and Defend, is a long-lasting soil fungicide.
- It controls various soilborne diseases of vegetables, turf, and ornamentals and is applied as a dip or in the furrow at planting time.
- It is used primarily against Rhizoctonia and Plasmodiophora.
- **Dichloran (DCNA)**

- Sold as Botran and Allisan,
- is used as foliar and fruit fungicide or postharvest spray for diseases of vegetables and flowers caused mostly by Botrytis, Sclerotinia, or Rhizopus
- **Chlorothalonil**
- Available as Bravo, Daconil, Exotherm Termil, and several other trade names.
- An excellent broad-spectrum fungicide against many leaf spots, blights, downy mildews, rusts, anthracnoses, scabs, and fruit rots of many vegetables, field crops, ornamentals, turf, and even trees.
- A tablet formulation of chlorothalonil called Termil is dispersed thermally in green houses for the control of Botrytis on many ornamentals and for several leaf molds and blights of tomato.
- Biphenyl, sold as biphenyl, has been used widely for control of postharvest diseases of citrus caused by Penicillium, Diplodia, Botrytis, and Phomopsis.
- Biphenyl is volatile and is applied by impregnating shipping materials with it; the compound then volatilizes in storage and protects the stored fruit.

Heterocyclic Compounds

- Heterogeneous group.
- Eg. Captan, Iprodione, and Vinclozolin.
- Most of them also inhibit the production of essential compounds containing —NH2 and —SH Groups (Amino compounds and enzymes).

1. Captan

- Excellent fungicide.
- Control of leaf spots, blights, and fruit rots on fruit crops, vegetables, ornamentals, and turf.
- Used as seed protectant for agronomic crops, vegetables, flowers, and grasses.
- Reported to repel "seed pulling" birds.

2. Iprodione

- Broad-spectrum, Foliage-contact fungicide.
- Inhibits spore germination and mycelial growth.
- Shows mostly preventative and only early curative activity.

- Effective against Botrytis, Monilinia, and Sclerotinia and also against Alternaria and Rhizoctonia.
- Applied most often as a foliar spray and also as a postharvest dip and as a seed treatment.
- Iprodione is used on turf, stone fruits, grapes, peanuts, onions, lettuce, and other crops.
- Used as a protective systemic and curative fungicide against the basidiomycetes Rhizoctonia, *Sclerotium rolfsii,* Corticium, and Typhula.
- It is applied as a spray.

3. Vinclozolin

- Contact protective fungicide.
- Effective against sclerotia-producing ascomycetes (Botrytis, Monilinia, Sclerotinia) and other fungi.
- Used mostly as a spray on strawberries, lettuce, turf, aornamentals, and on fruit.

8

Systemic Fungicides

Definition

Systemics fungicides are the fungicitoxic compounds which when applied on different parts of the plant are absorbed by the plant tissues and then translocated upwards, downwards and both ways and act on the pathogen either directly or through its metabolite products and control plant diseases away from the point of application.

Specific characters of systemic fungicides

1. Absorbed through the foliage or roots.
2. Translocated within the plant through the xylem.
3. Generally, move upward in the transpiration stream and may accumulate at the leaf margins.
4. A few of them, e.g., fosetyl-Al, also move downward.
5. Not re-exported to new growth.
6. Most are only locally systemic within the sprayed leaves.
7. Some of them become translocated systemically when sprayed on herbaceous plants.

Concept of systemic fungicides

- Fungicide when applied on the plant surface they may;
    - Remain on the surface: Non-systemic
    - Absorbed by the plant: Systemic
- Systemic fungicides may
    - Remain there in treated plant parts: Locosystemic
    - May move in the direction of evapotranspiration stream: Apoplastic
    - May move with photosynthates to sink: Symplastic
    - Or in both directions: Ambimobile.
- **Symplast** is the living part of the plant which is enclosed by membranes, i.e. protoplasts, and plasmodesmata, including the phloem sieve cells.

- Long distance transport in phloem is symplastic.
- **Apoplast** is the nonliving part of the plant, i.e., cell walls and cuticle, including xylem vessels and tracheids.
- Long distance movement in the transpiration stream is apoplastic.

Systemic Fungicides

1. Effective when applied as seed treatments, root dips, in-furrow treatments or soil drenches, and in trees when injected into the trunks.
2. Almost all systemic fungicides are site specific, inhibiting only one or perhaps a few specific steps in the metabolism of the fungi they control.
3. As a result, many target fungi through simple mutation and become resistant to each frequently used systemic fungicide within a few years of introduction of the compound.
4. For this reason, various strategies have been developed for preserving the usefulness of such chemicals.
5. To avoid abandonment of a systemic fungicide after appearance of a pathogen strain resistant to it, the fungicide must be used in combination with another broad-spectrum contact fungicide under various schemes of application.

Acylalanines

1. Most important acylalanine is the fungicide metalaxyl.
2. Effective against the oomycetes Pythium, Phytophthora, and several of the downy mildews.
3. It is sold as Ridomil , Apron, Subdue.
4. Metalaxyl is one of the best systemic fungicides against oomycetes. Pythium and Phytophthora certain downy mildews (e.g., of tobacco).
5. It is also effective as a curative treatment if it has to be applied after infection has begun.
6. Metalaxyl is quite water soluble and is translocated readily from roots to the aerial parts of most plants, but its lateral translocation is slight.
7. Because the use of metalaxyl has already resulted in the appearance of strains resistant to it in some pathogens, it is recommended that it be used in combination with other, broad-spectrum fungicides.

Benzimidazoles

1. Include Benomyl, Carbendazim, Thiabendazole, and Thiophanate.
2. Effective against numerous types of diseases caused by a wide variety of fungi.

3. Most benzimidazoles are converted at the plant surface to methyl benzimidazole carbamate (MBC, carbendazim).
4. This compound interferes with nuclear division of sensitive fungi.

I.BENOMYL

1. Safe, broad-spectrum fungicide, effective against a wide range of leaf spots and blotches, blights, rots, scabs, and seed-borne and soilborne diseases.
2. Particularly effective for powdery mildew of all crops; scab of apples, peaches, and pecans; brown rot of stone fruits; fruit rots in general; Cercospora leaf spots; cherry leaf spot; black spot of roses; blast of rice; and various Sclerotinia and Botrytis diseases.
3. It is highly active against and suppresses infection by Rhizoctonia, Thielaviopsis, Ceratocystis, Fusarium, and Verticillium.
4. It has no effect on oomycetes, on some dark-spored imperfect fungi such as Bipolaris, Drechslera, and Alternaria, on some Basidiomycetes, and on bacteria.
5. Benomyl may be applied as a seed treatment, foliar spray, trunk injection, root dip, or row treatment, and as a fruit dip.

II. THIABENDAZOLE

1. Sold as Mertect 340-F, Arbotect 20- S, and Decco Salt No.19.
2. Broad-spectrum fungicide and is effective against many imperfect fungi causing leaf spot diseases of turf and ornamentals anddiseases of bulbs and corms.
3. Commonly used as a postharvest treatment for the control of storage rots of citrus, apples, pears, bananas, potatoes, and squash.

III. THIOPHANATE

1. Under the trade name Topsin.
2. Effective against several root and foliage fungi affecting turf grasses and vegetable crops.
3. Thiophanate methyl, under the trade names Fungo, Topsin M, Domain, Cavalier, Halt, etc.,
4. Broadspectrum preventive and curative fungicide for use on turf and as a foliar spray to control powdery and downy mildews, Botrytis diseases, numerous leaf and fruit spots, scabs, and rots.
5. Also used as a soil drench or dry soil mix to control soilborne fungi attacking bedding plants, foliage plants, and container-grown plants.

Oxanthiins

1. Oxanthiins were the first fungicides to be discovered as having systemic activity (1966).
2. They include primarily carboxin and oxycarboxin and are effective against some smut and rust fungi and against Rhizoctonia.
3. Oxanthiins are selectively concentrated in cells of these fungi and inhibit succinic dehydrogenase, an enzyme important in mitochondrial respiration.
4. Carboxin is sold as Vitavax. It is used as a seed treatment and is effective against damping-off diseases caused by Rhizoctonia and against the various smuts of grain crops.
5. Oxycarboxin is sold as Plantvax or Carbojec.
6. It is sometimes used as a seed or foliar treatment and is effective in controlling a wide variety of rust diseases.
7. Flutolanil, is sold as Contrast, ProStar, and Moncut. Nicobifen, belonging to a new anilide family of fungicides, interferes with mitochondrial respiration and energy production, moves in a translaminar and acropetal systemic manner, and controls a range of Ascomycetes on many crops.

Organophosphate Fungicides

1. Include fosetyl-Al, sold as Aliette, kitazin (IBP), edifenphos (Hinosan), pyrazophos (Afugan), and phosphorous acid, sold as Fosphite.
2. Aliette is very effective against foliar, root, and stem diseases caused by oomycetes such as Phytophthora, Pythium, and downy mildews in a wide variety of crops.
3. It is applied as a foliar spray, soil drench, root dip, or postharvest dip, and in soil incorporation.
4. Treatments may be effective for 2 to 6 months, depending on the crop.
5. Fosetyl-Al has been reported to stimulate defense reactions and the synthesis of phytoalexins against oomycetes.
6. Kitazin (IBP) and edifenphos (Hinosan), both effective against rice blast and several other diseases,
7. Pyrazophos (Afugan), which is effective against powdery mildews and Bipolaris and Drechslera diseases on various crops.

Pyrimidines

1. Pyrimidines include diamethirimol (Milcurb), ethirimol (Milstem), and bupirimate (Nimrod), all effective against powdery mildews of various crop plants.

2. Fenarimol (Rubigan) and Nuarimol (Trimidal) are effective against powdery mildews and also several other leaf spot, rust, and smut fungi.

Triazoles

1. Triazoles (conazoles or imidazoles) include several excellent systemic fungicides, such as triadimefon (Bayleton), triadimenol (Baytan), bitertanol (Baycor), difenoconazole (Divident, Score), fenbuconazole or butrizol (Indar or Enable), propiconazole (Tilt, Orbic, Banner, Alamo), etaconazole or cyprodinil (Vangard), myclobutanil (Rally, Immunox, Spectracide- Pro, Nova, Eagle, Systhane), cyproconazole (Sentinel), and tebuconazole (Elite, Folicure, Raxil, Lynx).
2. They show long protective and curative activity.
3. Effective against a broad spectrum of foliar, root, and seedling diseases such as leaf spots, blights, powdery mildews, rusts, smuts, and others caused by many ascomycetes, imperfect fungi, and basidiomycetes.
4. They are applied as foliar sprays and as seed and soil treatments.

Strobilurins or QoI Fungicides

1. Contains the newest and most important fungicides.
2. The first such fungicide was isolated from the wood-rotting mushroom fungus *Strobilurus tenacellus*.
3. Thought to help the fungus defend itself from other microbes present in rotting wood.
4. Subsequently, chemists produced more effective and more stable strobillurin compounds.
5. Strobilurins have a common mode of action, i.e., all of them interfere with respiration, i.e., energy production, in the fungal cell.
6. They do that by blocking electron transfer at the site of quinol oxidation (the Qo site) in the cytochrome bc1 complex, thereby preventing ATP formation.
7. Strobilurins are, therefore, site-specific fungicides and as such are subject to selecting for fungicide-resistant strains of fungi and development of fungicide (strobilurin)-resistant pathogen populations.
8. Move translaminarly within leaves.
9. In addition, some strobilurins, such as azoxystrobins, move translaminarly and systemically through the vascular system of the plant.
10. Some strobilurin fungicides show growth-promoting effects on treated plants, apparently by delaying leaf senescence and having water-conserving effects.

11. Strobilurins are Azoxystrobin, sold as Abound, Heritage, or Quadris; Trifloxystrobin, sold as Flint for use on grapes, pome and stone fruits, cucurbits, fruiting vegetables, as Gem for use on citrus, rice, and sugar beets, and as Stratego for ruse on peanuts, wheat, corn and rice.
12. Kresoxim methyl, sold as Sovran and Cygnus. Pyraclostrobin, sold as Insignia.
13. Strobilurins are effective against most fungal diseases of most crops.
14. However, they seem to increase the severity of a few diseases, probably by eliminating or suppressing some naturally occurring microorganism(s) that is antagonistic to the pathogen.

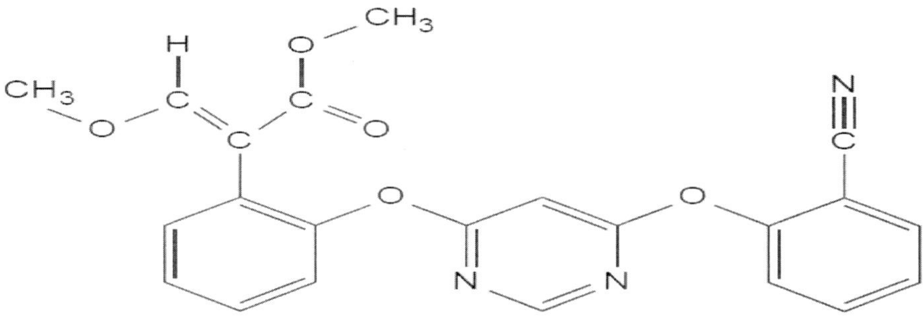

Azoxystrobin

Miscellaneous Systemics

Several excellent systemic fungicides of different chemical composition and affiliation are included in the miscellaneous category.

1. Chloroneb

- Sold as chloroneb.
- Used as a seed treatment for beans, soybeans, and cotton and as a soil fungicide for turf and ornamentals.
- Sometimes applied as a seed overcoat to seed treated with standard fungicides.
- It does not leach from the soil.

2. Ethazol

- Sold as Truban, Terrazole, or Koban.
- Seed, soil, and turf fungicide.
- Effective against dampingoff and root and stem rots caused by Pythium and Phytophthora.

- Often sold combined with PCNB or with thiophanate methyl (Banrot) for broader spectrum, application, particularly against Fusarium and Rhizoctonia.

3. Imazalil

- Sold as Fungaflor, Flo-Pro, or Nu-Zone.
- Effective against many ascomycetes and imperfect fungi causing powdery mildews, leaf spots, fruit rots, and vascular wilts.
- Applied as a foliar spray, a seed treatment, and as a postharvest treatment.
- Has excellent curative and preventive properties.

4. Prochloraz

- Also, an imidazole sold as Prochloraz
- Effective against ascomycetes and imperfects causing powdery mildews, leaf spots, and blights and fruit rots.
- It is used as a spray or a seed treatment.

5. Propamocarb

- Sold as Banol and Previcur
- Effective against Pythium, Phytophthora, downy mildews, some rusts, and others.
- Applied as a seedling dip, soil drench, seed treatment, soil surface spray, and foliar spray.

6. Triforine

- Sold as Funginex or Triforine,
- Effective against many ascomycetes and imperfect fungi causing powdery mildews, foliar and fruit spots, fruit rots, anthracnose, and some basidiomycetes causing rusts.
- It is used as a foliar spray.

Miscellaneous Organic Fungicides

Chemically diverse compounds are excellent protectant fungicides for certain diseases or groups of diseases.

Dodine

- Sold as Syllit.
- It is an excellent fungicide against apple scab and also controls certain foliage diseases of cherry, strawberry, pecan, and roses.
- It gives long-lasting protection and is also a good eradicant.

- It appears to have limited local systemic action in leaves.
- Strains of the apple scab fungus resistant to dodine have appeared and predominate in some areas.

Fentin hydroxide

- Sold as Super Tin,
- Broad spectrum fungicide with activity against many leaf spots, blights, and scabs. It also has suppressant or antifeeding properties on many insects.

Fludioxonil

- Used for different crops and postharvest uses.

Famoxadone

- Sold as Famoxate, is effective against Ascomycetes and Oomycetes on cucurbits, etc.

9

Modes of Action of Fungicides

Fungicides are metabolic inhibitors and their modes of action can be classified into four broad groups.

1. Inhibitors of electron transport chain.
2. Inhibitors of enzymes.
3. Inhibitors of nucleic acid metabolism and protein synthesis.
4. Inhibitors of sterol synthesis.

Inhibition of Electron Transport Chain

(Respiration in mitochondria)

1. **Sulfur**

- Disrupts electron transport along the cytochromes.

2. **Strobilurins**

- Inhibit mitochondrial respiration, blocking the cytochrome bc1 complex.
- Examples: Azoxystrobin, Kresoxim-methyl, Pyraclostrobin, Trifloxystrobin.

Inhibition of enzymes

1. Copper

- Nonspecific denaturation of proteins and enzymes.

2. Dithiocarbamates

- Inactivate –SH groups in amino acids, proteins and enzymes.
- Examples: Maneb, Zineb, Mancozeb, etc.

3. Substituted aromatics

- Inactivate amino acids, proteins and enzymes by combining with amino and thiol groups.
- Examples: Chlorothalonil, PCNB.

4. Organophosphonate

- Disrupts amino acid metabolism.
- Examples: Fosetyl-Al.

Inhibition of Nucleic Acid Metabolism & Protein Synthesis

1. Benzimidazoles

- Inhibit DNA synthesis (Nuclear division).
- Examples: Thiophanate-methyl, Carbendazim, Benomyl.

2. Phenylamides

- Inhibits RNA synthesis.
- Examples: Metalaxyl, Mefenoxam.

3. Dicarboximides

- Inhibits DNA and RNA synthesis, cell division and cellular metabolism.
- Examples: Iprodione, Vinclozolin

Inhibition of Sterol Synthesis

1. Imidazoles

- Imazalil.

2. Triazoles

- Propiconazole, Myclobutanil, Tebuconazole, Triflumazole, Azoxystrobin

3. Morpholines (Dimethomorph)

- Inhibits sterol production at different site than imidazoles and triazoles.
- Affects cell wall production.

Mode of action, targetsite and resistance risk attached to different chemical classes of fungicides

Fungicide class	Mode of action	Target site	Resistance risk
Anilino- pyrimidines	Aminoacids and protein synthesis	Methionine biosynthesis	Moderate
2-Aminopyrimidines	Nucleic acid synthesis	Adenosin- deaminase	Moderate
Benzimidazoles	Mitosis and cell divison	B – tubulin assembly	High
Carbamates	Lipids and membrane synthesis, fatty acid	Cell membrane	Moderate
Cinnamic acid derivatives	Glucan and cell wall synthesis	Cell wall synthesis	Moderate
Cyano-acetamide oximes	Nucleic acid/protein synthesis	Undefined	Moderate

Copper compounds, inorganic sulphur, dithiocarbamates, phthalimides, chlorophenuls	Multisite disruption of cell function	Reaction with enzymes having –SH amino and hydroxyl groups	Low
Carboxamides	Nucleic acid synthesis	Complex II : succinate dehydrogenase	Moderate
Demethylation inhibitors (DMIs)	Sterol biosynthesis inhibitors	C14- demethylase in sterol biosynthesis	Moderate
Dicarboxamides	Lipid and membrane synthesis	NADPH cytochrome c-reductase in lipid peroxidation	High
Morpholins	Sterol biosynthesis inhibitors	D 8-7 isomerase in sterol biosynthesis	Moderate
N-phenylcarbamates	Mitosis and cell divison	B – tubulin assebly	High
Phenylamides	Nucleic acid sythesis	RNA polymerase	High
Phenylpyrroles	Signal transduction	MAP proetein kinase	Moderate
Strobilurins (Q_oI fungicides)	Respiration	Cytochrome III: cytochrome be I(ubiquinol oxidase)	High

10

Formulations of Fungicides

Fungicide formulation is the process by which the fungicide is put into a form which can be easily produced, stored, transported and applied by practical methods in order to achieve a safe, convenient, economic and effective method of disease control.

Objective of Festicide Formulation

- To uniformly spread a small amount of the active ingredient over a large area.
- To ensure safety in handling and application.
- To optimize funicide doses for better efficacy.

Types of Fungicide Formulations

1. Solid formulation
2. Liquid formulation

Solid Formulations

Dusts (D)

- Free flowing powers contain tehnical material in the range of 2 to10% inert carrier.

Granules (G)

- Genuine products contain technical material in the range of 3 to 10% and granule base.

Water dispersible powders (WP)

- Free flowing powders containing technical material in the range of 25 to 75% wetting, dispersing agents and carrier.

Soluble powders (SP)

- Soluble powders provide most of the same benefits as wettable powders, without they need for agitation once they are dissolved in the spray tank.

Liquid formulation

1. Emulsifiable Concentrates (EC)

- Liquid formulations based on technical material which are not soluble in water contain 25 to 80% technical material solvent and emulsifier.

2. Water Soluble Liquids (SL)

- Liquid formulation based on technical material which are soluble in water contain 36 to 85% technical material and solvent.

3. Suspension Concentrates

- Thick slurry like formulations mixable in water.

4. Aerosols

- They are liquids under pressure filled incans which on release give a misty spray.

5. ULV Formulations

- These are liquid formulations suitable for ultra-low volume application.

11

Resistance to Fungicides

Definition

Resistance to fungicide may be defined as 'a heritable change in the sensitivity of a pathogen population that is reflected in the repeated failure of a fungicide to achieve the expected level of control when used according to the label recommendation for that pathogen.

Instances of resistance development to compounds of major classes of fungicides used in plant disease control

Fungicide group/compound	Main Pathogens affected	References
Benomyl	*Botrytis cinerea*	Grape
Carbendazim	*Venturia inaequalis*	Apple
Metalaxyl	*Pseudoperonospora cubensis*	Cucurbits
	Plasmopara viticola	Grapes
	Phytophthora infestans	Potato
	Bremia lactucae	Lettuce
	Botrytis cinerea	Strawberry
	Corynespora cassiicola	Cucumber
	Botrytis cinerea	Grapes
	Alternaria alternata	Oil seeds
	Venturia inaequalis	Apple
	Sclerospora fuliginea	Cucurbit and barley
	Mycosphaerella graminicola.	Cereal
Edifenphos	*Magnoporthe grisea*	Rice
Ethirimol	*Erysiphe graminis*	Barley
	Alternaria solani	Potato
	Colletotrichum graminicola	Barley
	Cercospora sojina	Soybean

Causes of Resistance Development

The causes and rate at which fungicide resistance develops depend on several factors, including the

- Initial frequency of resistance alleles present in the population.
- How rapidly the pathogen reproduce.
- Pathogens' level of resistance.
- Migration and host range of the pathogen.
- Fungicide persistence and specificity.
- Rate, timing and number.

Factors which encourage development of resistance

- Using the same fungicide, or fungicides from the same chemical group, too frequently in the same season.
- Using too high a dose of the fungicide.
- Using a fungicide with a long persistence.
- Higher proportion of resistant individuals in the population before the fungicide is used.
- Wide area treatments.
- A short pathogen life cycle.

Strategy for fungicides resistance management

1. Monitor pathogens.
2. Ecological engineering.
3. Take an integrated approach.
4. Mix and apply carefully.
5. Alternate different fungicides classes.
6. Preserve susceptible genes.
7. Use fungicides as last resort.

12

Antibiotics in Plant Desease Management

Definition

"Antibiotic is defined as a chemical substance produced by one microorganism, which in low concentration can inhibit or even kill other microorganism".

Character of good antibiotics

- Broad spectrum.
- Safe to plants.
- Active on or inside of the plant.
- Tolerates oxidation, UV irradiation, rainfall, and high temperature.
- Lead to development of resistant pathogens only at a low or non-detectable dose.

Advantages & Disadvantages in of using Antibiotics

Advantages of using antibiotics

- Easy to produce antibiotics at cheaper rate.
- Cure incurable diseases.
- Less number of sprays required.
- Less leaching effect.
- Acts as eradicant.
- Manage MLO's and bacterial diseases.

Disadvantages of using antibiotics

- Shelf life very short.
- Less stable in storage.
- Narrow site of action.
- Risk of development of resistance.

Classification of Antibiotics

Based on chemical group

1. Broad-spectrum antibiotic

- Effective against a large number of microorganisms.
- E. g. Tetracylines, Chloromycetin etc.

2. Narrow-spectrum antibiotic

- Effective against specific organisms only.
- E. g. Penicillin

Based on chemical group

1. β- Lacta antibiotics

Ex: a. Penicillins

b. Cephlosporins

2. Tetracyclins

Ex: a. Tetracyclins

b. Chlorotetracycline

c. Oxytetracyclins

3. Rifamycins

Ex: Rifamycin A to E

4. Amino glycosidic antibiotics

Ex: Streptomycin, Gentamycin

5. Macrolide antibiotics

a. Erythromycin

b. Spyramycine

c. Olendomycine

6. Polypeptide group antibiotics

Bacitracine, Capriomycine, Polymixin, Viomycine

7. Miscellaneous group

Vancomycine, Chloromphenical, Fusidic acid, Lincomycine

8. Synthetic antibiotics

Ex: Nitrofuron compounds

a. Furazolidin - Stomach

b. Nitrofurantoin - Urinary track infection

c. Nitrofurozone - Ear infection

Important antibiotics and their mode of action

Name	Source	Effective against	Mode of action
Penicillins	*Penicillium spp.*	Prokaryotes	Inhibits murein synthesis
Cephalosporins	*Cephalosporium spp.*	Prokaryotes	Inhibits murein synthesis
Streptomycin	*Streptomyces griseus*	Prokaryotes	Binds to protein S12 of 30S causing aberrant inhibiton complex
Kanamycin	*Streptomyces kanamyceticus*	Prokaryotes	Affect 30S risbosomal aubunits and prevents the translation.
Neomycin	*Streptomyces fradiae*	Prokaryotes	Affect 30S risbosomal aubunits and prevents the translation.
Erythromycin	*Streptomyces erythraeus*	Prokaryotes	Inhibits translation by binding to the 50S ribosome
Carbomycin	*Streptomyces halstedii*	Prokaryotes	Inhibits translation by binding to the 50S ribosome
Chlorotetracyclin	*Streptomyces aureofaciens*	Prokaryotes	Inhibits binding of aminoacyl-t- RNA to A site of 30S ribossomes; specificallybinds to the protein S& near A site, resulting change in topology
Oxytetracycline	*Streptomyces riomosus*	Prokaryotes	Inhibits binding of aminoacyl-t- RNA to A site of 30S ribossomes; specificallybinds to the protein S& near A site, resulting change in topology
Rifampin	*Amycolatopsis rifamycinica*	Prokaryotes	Inhibits DNA dependent RNA polymerase in bacterial cells by binding its betasubunit, thus preventing transcription to RNA and subsequent translation to proteins.
Bacitracin	*Bacillus subtilis*	Prokaryotes	Destroys murein biosynthesis.
Polymixin-G	*Bacilus polymyxa*	Prokaryotes	Destroys cytoplasmic membrane.

Chloramphenicol	*Streptomyces venezuelae*	Prokaryotes	Inhibits peptidyl transferase activity of 50S ribosome and affects translation.
Nystacin	*Streptomyces nouresii*	Prokaryotes	Inactivates membrane containing sterols by making pore through which kions , small molecules.
Amphotericin	*Streptomyces nodosus*	Prokaryotes	Inactivates membrane containing sterols by making pore through which kions , small molecules.
Aureofungin	*Streptoverticillium cinnamoneus var terricola*	Eukaryotes (Fungi)	Inactivates membrane containing sterols by making pore through which kions , small molecules.
Natamycin/ Pimaricin	*Streptomyces natalensis*	Prokaryotes	Inactivates membrane containing sterols by making pore through which kions , small molecules.

13

Acquaintance with Plant Protection Equipments

Spraying and Dusting Equipments

Appliances that are used for applying dust and spray formulation of pesticides are called dusters and sprayers, respectively.

Uses of Spraying and Dusting Equipments

- Applying fungicides and bactericides to control the plant diseases.
- Applying insecticides to control insect pests on crops and in stores, houses, kitchens, poultry farms, barns, etc.
- Applying acarices to control phytophagous mites.
- Applying herbicides, to kill the weeds.
- Applying harmone sprays application to increase the fruit set or to prevent the premature dropping of fruits.
- Applying plant nutrients as foliar spray.
- Applying powdery formulation of poisonous chemicals on the crops and for any other purposes.

Sprayers

A **sprayer** is a device used to spray a liquid.

Types of sprayers

- Hand sprayer
- Knapsack sprayer
- Rocker sprayer
- Bucket sprayer
- Foot sprayer
- Power sprayer

Hand sprayer

- Used for spraying small areas like kitchen, garden and experimental laboratory plots.

- Capacity of the container - 500 to 1000 ml.
- Hydraulic energy sprayer.
- It has a hydraulic pump inside the container, with cylinder, plunger and a plunger rod.
- By operating the plunger up, the spray fluid in the container is sucked into the cylinder through a ball valve assembly and then pressurized during the downward stroke.
- The pressurized fluid is then let out through a nozzle, and sprayed into fine droplets.

Knapsack sprayer

- Any sprayer which is carried on the back of the operator is called a knapsack sprayer.
- Commonly used manually operated knapsack sprayer will have one hydraulic pump working inside the container.
- Plunger works inside the replacement well attached at the bottom of the container, for easier maintenance.
- Pump can be operated through the appropriate linkages by oscillating the handle, with the sprayer carried on the back.
- An agitator is also provided with the pressure chamber to agitate the fluid so that the particles in suspension will not be allowed to settle down.
- A delivery tube is attached on the other end of the pump which carries the pressurizes fluid to the spray lance.
- Flow to the nozzle is controlled by a trigger cut-off valve.
- Unit is again brought back to the ground for pumping air and then the spraying is contained as before.
- Spray fluid, which does not enquire any agitation only can be sprayed by using this type of sprayers.

Rocker sprayer

- The rocking sprayer has a pump assembly, fixed on a wooden platform with an operating lever, a valve assembly with two ball valves, a pressure chamber, suction hose with strainer, and delivery hose with spray lance.
- When the plunger is pulled behind by pulling the lever way from the pump, the spray fluid from the container is sucked through the strainer and pushes the bottom ball valve above and enters the pump.
- The movement of the lower ball valve is arrested by the upper valve seat.

- When the lever is pushed towards the pump, the sucked fluid is forced to enter the pressure chamber by opening the upper ball valve.
- The operation is continued till the entire suction pipe, ball valve assembly, delivery hose and a portion of pressure vessel is fitted with spray fluid and the pump operator finds it difficult to push the piston forward, due to the downward pressure developed by the entrapped compressed air in the pressure vessel.
- Thereafter, the trigger cut off valve will be opened to allow the spray fluid to rush through the nozzle and get atomized.
- Usually 14 to 18 kg/cm2 pressure can be built in the pressure chamber and hence can be conveniently used for free spraying.

Bucket sprayer

- Designed to pump the spray fluid directly from, the open container, usually a bucket.
- Hydraulic pump will be put inside the bucked and held properly with the help of foot rest.
- As the plunger is pulled up, the fluid enters through the suction ball valve assembly and when the plunger is pressed down, the suction valve closes and the fluid enters the pressure chamber through a ball valve assembly.
- As the plunger is continuously worked, pressure is built in the pressure chamber and the delivery hose.
- As soon as the required pressure is built up, the spraying will be done.
- A pressure of 4 kg / cm2 is developed in most of the models.

Foot sprayer

- Modified version of rocker sprayer.
- Pump is fixed in a vertical position with necessary braces.
- Plunger moves up and down when operated by the pedal.
- A ball valve is provided in the plunger assembly itself to allow the fluid to cross the plunger and getting pressurized in the pressure vessel.
- During the upward motion of the piston fluid is sucked in and pressurized into the pressure vessel and during downward movement, the sucked fluid crosses the plungers and enter the pump.
- Pressure developed is about 17-21 kg/cm2.

Power sprayer

- All the sprayers which impart the mechanical energy developed by an I.C. Engine, on the spray fluid before spraying is called as a power sprayer.

- The most commonly used type of power sprayer in India is a gaseous energy type knapsack sprayer.
- In construction, it has a back pack stand on which a blower with a S.I. Engine of 1.2 to 3 hp capacity, the spray fluid tank and the petrol tank are fixed rigidly.
- A pleated hose is attached to the blower elbow to carry the high velocity air and at the end of that a shear nozzle is fixed to allow the spray fluid to trickle in from the spray fluid storage tank, with a valve control.
- From the top of the blower casing, an air hose is taken into the spray fluid tank, which carries little quantum of air to press the spray fluid during operation.
- In operation, the engine is started by keeping the unit on the ground and then carried by the operator.
- Blower sucks the air behind the backrest and forces it into the pleated hose.
- Valve of the shear nozzle is opened or the shear nozzle with selective opening and discharged through the nozzle.
- High velocity air shears off the droplets and atomizes by the impact of diffuse and delivers it on the plant the surface.
- An air current of 2.7 to 9.1 m2 / minute is delivered at a velocity of 175 to 320 kmph.
- Spray fluid tank capacity varies from 7 to 12 litres.
- Fuel tank capacity varies from 0.75 to 2.25 litres.
- Spray fluid discharge can be varied from 0.5 to 5 lit / minute.

Dusters –

The mechanical appliances that are used for distributing the dust formulations of pesticides are called as dusters.

Types of dusters

- Package duster
- Plunger duster
- Bellow duster
- Rotary duster

Package dusters

- In some pesticide dusts are packed in containers that serve as a hand applicators and may be discard after use.

- They are mostly provided with rubber, leather or plastic section which, on getting squeezed, provides a puff of air that emits the dust in a small cloud.
- Simplest type of package duster is worked by pressing it between the fingers.

Plunger dusters

- Simple in construction and consists of a dust chamber, a cylinder with a piston or plunger, a rod and a handle.
- Useful for small scale use in kitchen garden and in household.

Bellow duster

- In the below may be made from rubber, leather or plastic.
- On squeezing, it puffs the air that expels the dust in a small cloud.
- Hand held bellow duster has containers of capacity from 30 g to 500 g.
- Bellows can be operated either directly by hand or by handle provided for that purpose.
- Knapsack duster has the container capacity of 2.5 to 5.0 kg.
- Air blast developed by the bellow draws the dust from the hopper and discharges through the delivery spout intermittently.
- These dusters are suitable for spot treatments.

Hand Rotary duster

- Called crank dusters and fan type dusters.
- They may be shoulder mounted, back or belly mounted.
- Basically, a rotary duster consists of a blower complete with gear box and a hopper with a capacity of about 4-5 kg of dust.
- Duster is operated by rotating a crank and the motion is transmitted through the gear to the blower.
- Air current produced by the blower draws the dust from the hopper and discharges out through the delivery tube which may have one or two nozzles.
- Used for dusting field crops, vegetables and small trees and bushes in orchards.
- Efficiency of these dusters is 1 to 1.5 ha/day.

Power dusters

- Resemble the rotary duster is construction, except that the power to drive the blower through the gear box is tapped from an external power source which may be an engine or P.T.O. shaft of the tractor or flywheel of the power tiller.

- Power operated centrifugal energy knapsack sprayer also can be converted into a power duster, by allowing the dust fluid into the air stream, near the point of attaching the pleated hose, in the blower elbow.

Application Technique

Spray Volume: Depending upon the volume of spray required per hectare the application is categorized.

HV : High Volume = More than 150 l/ha.

- Suitable for insecticides, fungicides, herbicides.
- Can be done with knapsack sprayers, tractor mounted sprayers.

LV : Low Volume = Approx. 10-150 l/ha.

- Suitable for insecticides, fungicides.
- Can be done with motorized knapsack sprayer, air craft, (in other countries) low r.p.m. Spinning disc appliances are also used.
- Various equipment and techniques being used in last three decades are shown diagrammatically alongside.

ULV : Ultra Low Volume = Approx. 1-5 l/ha.

- Suitable for insecticides.
- Can be done with high r.p.m. spinning disc appliances, motorized knapsack sprayer fited with special spinning disc with special spinning disc attachment and air craft.

Nozzles

The nozzles perform four basic functions

- Atomizes liquid into droplets.
- Disperses the droplets in a specific pattern.
- Meters liquid at a certain flow rate.
- Provides hydraulic momentum.

Types of nozzle on the basis of suitability

Type of nozzle	Fungicide(3 bar)	Insecticide (3 bar)	Herbicide(1bar)
Hollow Cone	☼☼☼	☼☼☼	☼
Flat Fan	☼☼	☼☼	☼☼
Flood Jet	☼☼☼	☼	☼

☼ Unsuitable ☼☼ Acceptible ☼☼☼ Most suitable

Type of nozzle on the basis of size of droplets

Hollow Cone Pattern	Flat Fan Pattern.
• Produces finer droplets. • Spray penetrates better into the plant foliage. • Suitable for insecticide spray. • Spray angle is narrow to medium.	• Produces medium size droplets. • Give even and uniform coverage. • Suitable for herbicide spray. • Spray angle is narrow to medium.
Double Swirl Spray Nozzle • Used for spraying in two different directions simultaneously. • Nozzles can be fitted with different types of tips like hollow cone, solid cone or flat fan. • Suitable for high volume applications.	**Adjustable Nozzle.** • Most suitable for spraying targets which are not within the reach of man. • Gives a wide angle hollow cone to a straight solid stream that is, it gives a jet to a cone type of spray pattern. • Difficult to calibrate as the flow and droplet sizes vary widely with the nozzle angle.
Flood Jet Pattern • Produces coarse droplets gives wider coverage. • Suitable for herbicide spray. • Spray angle is wide.	

14

Plant Quarantine

Biosecurity

Biosecurity is strategic and integrated approach encompasses policy frameworks to analyze and manage risks for food safety, animal life and health, and plant life and health including environment.

Basic Elements

- Introduction of exotic pests/diseases.
- Emergence of virulent strains/races/ biotypes of economically important indigenous pests/diseases.
- Deliberate misuse of pests/pathogens.
- Biosafety issues arising from LMOs.

Plant Quarantine

- Term 'Quarantine' is derived from the Latin word quarantum, meaning 40.
- It refers to the 40 days period of detention of ships from countries with bubonic plague and cholera in the Middle Ages.
- First such quarantine was imposed in Venice in 1374.
- Plant quarantine , restrict entry of plants, plant products, soil, cultures of living organisms, packing materials, and commodities, as well as their containers and means of conveyance to protect agriculture and the environment from avoidable damage by hazardous organisms.

Goal of Plant Quarantine

- To delay or prevent entry of the pathogen along manmade pathways.
- If entry succeeds, to prevent infection.
- If infection succeeds, to prevent establishment of pathogens entering on manmade and natural pathways.
- If established, to minimize or retard spread.

Pests Introduced in India

Pests	Sicentific Name	Year of Intorudction
Lantana weed	*Lantana camara*	1809
Coffee rust	*Hemilia vastatrix*	1879
Late blight of potato	*Phytophthora infestans*	1883
Wooly apple aphid	*Eriosoma lanigerum*	1889
Grape downy mildew	*Plasmopara viticola*	1910
San Jose Scale	*Quadrapidiotus perniciosus*	1911
Water hyacinth	*Eichhornia crassies*	1914
Potato tuber moth	*Phthorimaea operculella*	1937
Banana bunchy top virus		1940
Wart of potato	*Synchytrium endobioticum*	1953
Congress grass	*Parthenium hysterophorus*	1956
Golden nematodes	*Globodera rostochiensis & G pallida*	1961
Apple scab	*Venturia inaequalis*	1978
Downy mildew of sunflower	*Plasmopara halstedii*	1984
Peanut stripe virus		1987
Sunflower necrosis	*Sunflower necrosis virus*	1997
Cotton mealy bug	*Phenococcus solenopsis*	2006
Papaya mealy bug	*Paracoccus marginatus*	2009
Tomato leaf minor	*Tuta abosoluta meyrick*	2014
Root knot nematode	*Meloidogyne enterolobii*	2016
White rust of chrysanthemum	*Puccinia horiana*	2016
Fall armyworm in maize	*Spodoptera frugiperda*	2018

Prerequisites of Plant Quarantine Regulations

- Only pathogens and pest that pose a threat to major crops or forest should be taken into considerations.
- It should be formulated to control the pests and not to hinder trade or accomplishment of other objectives.
- It should be operated solely under the law.
- Persons involved in quarantine measures should be properly trained and experienced.
- Professional workers and the public must co-operate on an international scale for effective operation of quarantine regulations.

Component of Plant Quarantine

- Exclusion of pathogens and pests of quarantine and economic significance.
- Suppression and eradication of exotic pathogens and pests recently introduced along natural and manmade pathways.
- Assistance to exporters of plant products in meeting the quarantine requirements of importing countries.

Pest Risk Analysis

- Pest risk defines the chances that a pathogen or pest of quarantine significance will enter along a manmade pathway.
- Risk often is expressed as low, medium or intermediate, and high.
    - **Low risk:** means that there is little chance that the pathogen or pest will enter.
    - **High risk:** means that chances are high that the pathogen or pest could enter.
    - **Acceptable risk:** An acceptable risk level means that the benefits derived after taking risk are high enough to justify taking the risk, with safe guards , in the first place.

Plant Quarantine Methods

1. Complete embargoes

- To exclude specified plants and plant products completely from a country infected or infested with highly destructive diseases or pests.
- However, in practice it is difficult to achieve because of more and more exchange of diverse genetic material among countries.

2. Partial embargoes

- Apply when a pest or disease of quarantine importance to an importing country is known to occur only in well-defined area of the exporting country.
- An effectively operating internal plant quarantine service exists that is able to contain the pest or disease within this area.

3. Inspection and treatment and certification at point of origin

- It involves the inspection and treatment of a given commodity when it originates from a country.
- Certification in accordance with quarantine requirements of importing country.

4. Inspection at the point of entry

- It involves inspection of plant material immediately upon arrival at the prescribed port of entry .
- If necessary subject to treatment before the same related.

5. Utilization of post entry plant quarantine facilities

- It involves growing of introduced plant propagating material under isolated or confined conditions.

Plant Quarantine in India

Sl .No	National legislations
1	Destructive Insects and Pests (DIP) Act of 1914
2	Plant Quarantine (Regulations of Import into India) Order 2003
3	Livestock Importation Act1898
4	The Environmental (Protection) Act 1986
5	Biological diversity Act 2002
6	Pesticide Management Bill, 2020

Pesticide Management Bill, 2020

- Pesticides Management Bill, 2020 was approved by the Union Cabinet on 12 February 2020.
- It replaces the existing Insecticide Act, 1968, which Prakash Javadekar, the Minister for Information and Broadcasting, mentioned as "age-old and needs immediate rewriting.

Aim of the Pesticides Management Bill

- Regulate the import, manufacture, export, storage, sale, transport, distribution, quality and use of pesticides.
- Control pests.
- Ensure availability of quality pesticides.
- Allow its use only after assessing its efficacy and safety.
- Minimize the contamination of agricultural commodities by pesticide residues.
- Create awareness among users regarding safe and judicious use of pesticides.
- To take necessary measures to continue, restrict or prohibit the use of pesticides with a view to prevent its risk on human beings, animals or environment.

Domestic Quarantine

Pest/Diseases	Host plant material	Restricted States	
		From	To
Wart (*Synchytrium endobioticum*)	Potato	Darjelling West Bengal	Any other state or place in India
Bunchy top (virus)of banana	Banana planting material	Assam, Kerala, Orissa, Tamil Nadu &West Bengal	Any other state or place in India & Uninon Territory
Mosaic (virus) of banana	Banana plants & planting material	Maharashtra &Gujarat	Any other state or place in India & Uninon Territory
Apple scab (*Venturia inaequalis*)	Apple planting materials	J&K and Himachal Pradesh	Any other states of India
Potato Cyst Nematodes (*Globodera rostochiensis &G pallida*)	Potato	Tamil Nadu	Any other state or place in India & Uninon Territory
Codling moth (*Carpocapsa pomenella*)	Apple and Walnut plants including fruits	Ladakh districts	Any other area of Jammu & Kashmir
Fluted scale (*Icerya purchase*)	Many host plant	Mysore (Karnataka), Madras (Tamil Nadu) &Kerala	Any other state or place in India
Coffee berry borer (*Hypothenamus hampei*)	Coffee seeds/ plants/powder	Nilgiri (TN), Kodagu (Karnataka) & Wynad (Kerala)	Any other states of India
San Jose Scale (Aspidiotus perniciosus)	Many host plant species	Punjab, UP, Madras (Tamil Nadu) & Kerala, WB, Assam, Orissa, HP, J&K	All other part of India

15

Disease Management through Use of Transgenic Plants

- Modern DNA technology made it possible to engineer transgenic plants.
- They are transformed with genes for
  - Resistance against specific diseases.
  - Coding for enzymes such as chitinases and glucanases directed against certain groups of pathogens like fungi, viruses, and bacteria.
  - Nucleic acid sequences that lead to gene silencing of pathogens.
  - Tolerance of adverse environmental factors.

Techniques of transformation of transgenic plants

1. Transgenic plants transformed with specific plant genes for resistance.
2. Transgenic plants transformed with genes coding for antipathogen.
3. Transgenic plants transformed with nucleic acids that lead to resistance and to silencing of pathogen genes.
4. Transgenic plants transformed with combinations of resistance genes.
5. Transgenic plants producing antibodies against the pathogens.

1. Transgenic Plants Transformed with Specific Plant Genes for Resistance

- There are numerous crops in which plant genes for specific pathogens have been isolated from resistant plants, transferred into susceptible plants, and expressed in these plants.
- All the necessary supporting genes are also transferred and expressed in the new host.
- Some of the formerly susceptible plants now behave as resistant ones.
- **Examples of crops transformed with plant resistance genes**

  Hybrid rice transformed with the rice gene **Xa21** coding for resistance of rice to the bacterial blight caused by *Xanthomonas oryzae pv. oryzae.*

- When the resistance gene DRR206 from pea was transferred into canola, the transgenic canola plants exhibited resistance to blackleg disease, caused by the fungus *Leptosphaeria maculans*.
- It decreased seedling mortality caused by the root pathogen *Rhizoctonia solani*, and resulted in smaller leaf lesions caused by *Sclerotinia sclerotiorum*.
- Similarly, in creeping bent grass plants transformed with the *Arabidopsis thaliana* gene PR5K, which codes for a protein kinase receptor whose extracellular domain is similar to the pathogenesis related (PR) proteins of the PR5 family.
- Transgenic plants showed resistance to the fungus *Sclerotinia homeocarpa* causing the dollar spot disease, with the resistance appearing as a delay in disease symptoms from 29 to 45 days.
- When tobacco and many other plants were transformed with animal antiapoptotic genes, the plants became resistant to necrotrophic pathogens, whereas transgenic plants in which this gene was destroyed were not protected against the pathogens.

Transgenic Plants Transformed with Genes Coding for Antipathogen

- Compounds genes coding for several pathogenesis-related (PR) proteins, such as chitinase and some glucanases.
- They are isolated, cloned, and expressed in plants, thereby interfering with the development of certain groups of pathogens and providing resistance to affected plants.
- **Examples of plants transformed with genes coding for antipathogen compounds include**
  - Peanut plants transformed with antifungal genes that reduced the incidence of Sclerotinia blight, caused by *Sclerotinia minor.*
  - Transgenic rose plants expressing a chitinase transgene from rice showed decrease in symptoms caused by the fungus *Diplocarpon rosae*, the cause of rose blackspot disease.
  - Transgenic broccoli plants expressing an endochitinase gene obtained from the biocontrol fungus *Trichoderma harzianum* had 14–200 times the endochitinase activity of that of controls and showed significantly less severe symptoms than nontransgenic plants.
  - Transgenic cotton and tobacco plants expressing the glucose oxidase gene obtained from the biocontrol fungus *Talaromyces flavus* showed significant resistance of seedlings to the Rhizoctonia fungus and partial resistance to Verticillium, but not to Fusarium.

- The glucose oxidase generates hydrogen peroxide, which, unfortunately, is toxic to both pathogens and plants.

- Transgenic *A. thaliana* plants expressing one or both of the genes coding for a cysteine protein inhibitor or, to a smaller extent, a cowpea trypsin (a serine) protein inhibitor, protect plants significantly from infection by several types of nematodes, especially the reniform nematode *Rotyenchus reniformis*.
- Increases in resistance shown for transgenic tobacco plants overexpressing a glutamate decarboxylase gene, which makes the plants resistant to the root knot nematode.
- Transgenic tobacco and potato expressing the bacterial gene ubiC, which leads to accumulation of toxic 4-hydroxybenzoic acid glucosides.
- Canola (*Brassica napus*) plants expressing an antimicrobial peptide, which makes the plants resistant to the blackleg disease caused by the fungus *L. maculans.*
- Peanut plants transformed with genes coding for antifungal enzymes, which make plants resistant to *S. sclerotiorum.*

Transgenic plants transformed with nucleic acids that lead to resistance and to silencing of pathogen genes

- Inserting segments of viral or other nucleic acids into plant genomes often leads to silencing of genes of the virus or subsequent pathogens that have homologous sequences, thereby making the plants resistant.
- There are several well-documented cases of successful transformation of a susceptible into a resistant crop through genetic engineering with parts of the genome of the virus.
- First such case of tobacco transformed with the coat protein gene of tobacco mosaic virus, which then became resistant to that virus.
- **For example,**
- **Insertion of a nontranslatable coat protein**
    - Insert a nontranslatable coat protein coding sequence of the tobacco etc virus (TEV).
    - Produces transgenic plants that develop symptoms on inoculated leaves.
    - However, rest of the plant remains free of symptoms.
- **Insertion of a gene for a double stranded RNase**
    - Insert a gene for a double stranded RNase from a yeast into the genome of pea plants.
    - It made the transgenic pea plants resistant to multiple viruses.

- **Transformed with the coat protein genome**
  - Squash transformed with the coat protein genome of cucumber mosaic, squash mosaic, and watermelon mosaic viruses.
  - Papaya transformed with the coat protein gene of papaya ring spot virus.
  - Soybean transformed with the CP nucleic acid of soybean mosaic virus.
  - Tobacco with the tobacco vein mottling virus.
  - Cantaloupe with the CP nucleic acids of cucumber mosaic virus, zucchini yellow mosaic virus, and watermelon mosaic virus-2, Potato with potato leaf roll virus; tomato to cucumber mosaic virus; squash to squash mosaic virus; citrus to citrus tristeza virus; chili peppers to cucumber mosaic virus and tobacco mosaic virus; and oilseed rape to turnip mosaic virus.
- **Viral nucleic acid used to transform the plant**
- **For example**
  - Peanut plants transformed and made resistant to the tomato spotted wilt virus by transferring into them an antisense nucleocapsid gene sequence.
  - Transgenic potato plants made highly resistant to potato virus Y by incorporating an antisense orientation of its P1 gene in the potato genome.
- **Viral replicase genome**
  - Transformed by introduction into their genome of the viral replicase gene.
  - Example, in transgenic wheat infected with wheat streak mosaic virus, in potato with potato leafroll virus, Tomato yellow leaf curl virus and cucumber mosaic virus in tomato, potato leafroll virus and potato virus Y in potato.
- **Movement protein genes**
  - Use of the movement protein genes of raspberry bushy dwarf virus in raspberry, tobacco mosaic virus in tomato, cymbidium mosaic virus in dendrobium.
  - Genes coding for viral inclusions in plant cells.
  - Genes coding for a variety of other nonstructural proteins of viruses.

- **Nonviral genes**
- Transformation of plants such as wheat, potato, pea, tobacco, and walnut with nonviral genes successful in making the transgenic plants resistant to several viruses.
- Some of the nonviral genes that have resulted in virus resistance include the gene for double stranded RNase (dsRNase) from *Schizosaccharomyces pombe.*
- Tobacco resistance gene N, mouse protein kinase, tobacco systemic-acquired resistance gene 8.2, and others.
- **Defective interfering DNAs or RNAs**
  - Resistance to viruses is mediated by small, defective interfering DNAs or RNAs.
  - Occur naturally in plant cells or are produced after inoculation with a DNA or RNA virus or transgene.
  - These small DNAs and RNAs are responsible for virus silencing as well as silencing of other genes of infected cells.
  - Generally, a transgene is supposed to be stable.
- **Inactivation of transgene activity can occur either**
  - A lack of transcription, probably as a result of DNA methylation of promoter regions and condensation of chromatin.
  - By instability of the transcripts (posttranscriptional silencing) as a result of initiation of a RNA degradation process affecting transgene RNAs and RNAs homologous to it.

Transgenic plants transformed with combinations of resistance genes

- Broad and effective resistance in many host/pathogen combinations are provided by combining a host gene for resistance with pathogen derived defense genes or with genes coding for antimicrobial compounds provides for a resistance.
- Transgenic rice plants carrying a host gene plus a promoter and first intron of the maize ubiquitin gene provide resistant to the rice blast (*Magnaporthe grisea*) disease.
- Combination of a tobacco host gene and a tobacco vein mottling virus coat protein gene, which showed broad and effective resistance to poty viruses in tobacco.
- Combination of the Sw5 tomato gene for resistance to tomato spotted wilt virus (TSWV) and the nucleocapsid (N) protein gene of TSWV in

transgenic plants, the latter showed high levels of resistance to several but not all TSWV strains.

Transgenic plants producing antibodies against the pathogens

- Plants lack an antibody-making machinery.
- But DNA technology made it possible to transform plants with additional genes that make possible the production of functional recombinant antibodies.
- Such plant-produced antibodies called plantibodies can be full antibody molecules.
- Fab fragments, or single chain Fv (scFv) fragments produced against specific plant viruses and they are targeted against the same viruses.
- scFv fragments are expressed in intracellular spaces, which allows them to be more effective in improving the resistance of the plant against the specific virus.
- Several plant viruses have been shown to be suppressed in transgenic plants transformed with genes.
- It enable them to produce full antibodies or fragments of single-chain variable regions of antibodies.
- Such viruses include tobacco mosaic virus, potato virus X, potato virus Y, and clover yellow vein virus.
- More work is needed before this method of plant disease control becomes truly effective and widely practiced.

16

Intergrated Disease Management

- A comprehensive approach to disease control that uses combined means to reduce the status of diseases to tolerable levels while maintaining a quality environment.
- The optimization of disease control in an economically and ecologically sound manner, accomplished by the coordinated use of multiple tactics to assure stable crop production and to maintain disease loss below the economic injury level while minimizing risks to man and the environment.

Goals of IDM

- Eliminate or reduce the initial inoculums.
- Reduce the effectiveness of initial inoculums.
- Increase the resistance of the host.
- Delay the onset of the disease.
- Slow the secondary cycles.

Objective of IDM

- To reduce potential hazards to environment and people health.
- To provides improved control by conservation of natural enemies and employing traditional method.
- To helps in the production of improved quality produce.
- To help to reduce the cost of plant protection.
- To helps in improvement on water and soil quality.
- To reduces farmers and consumer risk.
- To facilitates better pesticide management.

Why practice IDM

- Helps to keep a balanced ecosystem:
- Make pesticides more effective
- Can save money

- Promotes a healthy environment.
- Maintains a good public image.

Component of IDM

1. Acceptable disease levels
2. Monitoring
3. Genetic host resistance
4. Cultural practices
5. Mechanical controls
6. Biological controls
7. Chemical controls

Use of Disease Level Thresholds

Economic Injury Level: It is the lowest disease index that will cause economic damage.

Economic Threshold Level: It is the disease index at which control measures should be applied to prevent an increasing disease from reaching the economic injury level.

Action Threshold Level: An action threshold is the disease level at which some type of disease management action must be taken. It is a predetermined disease level that is deemed to be unacceptable.

Advantages of IDM

- Promotes sound structures and healthy plants.
- Promotes the sustainable bio based disease management.
- Reduces the environmental risk associated with management.
- Encourage adoption of more ecologically benign control tactics.
- Reduces the potential for air and ground water contamination.
- Protects the non-target species.
- Reduces the need for pesticides and fungicides by using several management methods.
- Reduces or eliminates issues related to pesticide residue.
- Decreases workers, tenants and public exposure to chemicals.
- Maintains or increases the cost-effectiveness of disease management programs.

Constraints of IDM

- There is a lack of multiple resistant varieties.
- There is a need for a number of selective pesticides against diseases of crops which are safe against the natural enemies.
- Economic threshold levels have not worked out for most of the diseases.
- Precise disease surveillance and monitoring methods and forecasting models have not been standardized.
- Techniques of mass multiplication of several potential bio control agents are still not well developed.
- Lack of multidisciplinary approaches inter institutional collaboration to develop sound IDM technology.
- Lack of trained human resource development in IDM.